About Island Press

Since 1984, the nonprofit organization Island Press has been stimulating, shaping, and communicating ideas that are essential for solving environmental problems worldwide. With more than 1,000 titles in print and some 30 new releases each year, we are the nation's leading publisher on environmental issues. We identify innovative thinkers and emerging trends in the environmental field. We work with world-renowned experts and authors to develop cross-disciplinary solutions to environmental challenges.

Island Press designs and executes educational campaigns, in conjunction with our authors, to communicate their critical messages in print, in person, and online using the latest technologies, innovative programs, and the media. Our goal is to reach targeted audiences—scientists, policy makers, environmental advocates, urban planners, the media, and concerned citizens—with information that can be used to create the framework for long-term ecological health and human well-being.

Island Press gratefully acknowledges major support from The Bobolink Foundation, Caldera Foundation, The Curtis and Edith Munson Foundation, The Forrest C. and Frances H. Lattner Foundation, The JPB Foundation, The Kresge Foundation, The Summit Charitable Foundation, Inc., and many other generous organizations and individuals.

The opinions expressed in this book are those of the author(s) and do not necessarily reflect the views of our supporters.

Purified

Purified

HOW RECYCLED SEWAGE IS
TRANSFORMING OUR WATER

Peter Annin

ISLANDPRESS | Washington | Covelo

Library of Congress Control Number: 2023938136

All Island Press books are printed on environmentally responsible materials.

Manufactured in the United States of America
10 9 8 7 6 5 4 3 2 1

Keywords: climate change, Colorado River, dead pool, de facto indirect potable reuse, direct potable reuse, drought, Hyperion 2035, indirect potable reuse, Operation NEXT, Orange County Water District, Pure Water San Diego, Pure Water Southern California, purified effluent, purified sewage, purified wastewater, purple pipe, sewage recycling, toilet to tap, water diversion, Water Factory 21, water purification, water recycling, water reuse, water scarcity

Contents

Author's Note

I HAVE BEEN WRITING ABOUT WATER FOR DECADES. From droughts in the Southwest to the Dead Zone in the Gulf of Mexico and water tension in the Great Lakes, I have seen firsthand how visceral, fraught, and life-changing water issues can be. But few water topics have fascinated me more than the one covered in this book: turning sewage into drinking water.

Like all things water, the history of purifying sewage is rife with controversy and dissent, but increasingly that dissent has morphed into acceptance, even enthusiasm. This shift is driven in part by utter desperation—large swaths of the United States are running out of water options. But it is also driven by an evolving faith in technology, especially among younger generations. As a result, water recycling is in the midst of a boom that stretches far beyond parched landscapes, and it is changing the national conversation about what we drink.

What surprised me most in reporting and researching *Purified* is how quickly water recycling is spreading around the country, yet most people are remarkably unaware of what's happening behind the scenes. This book is designed to fill that gap. As we leave what I call the century of oil and increasingly work our way into the century of water, it is incumbent upon all engaged citizens to understand and embrace the complex issues that define the water debate. The rapid ascent of the water recycling movement shows that in the climate change era, water cannot be taken for granted anymore—and that includes sewage.

—Peter Annin

"Do You Drink Beer?"

Greg Wetterau knows water. An expert on desalination and water recycling, he has helped build more than fifty high-tech water treatment plants around the world. His specialty is creating potable water by using "unconventional approaches," which means he's a master at turning saltwater—and sewage—into drinking water. He has overseen the construction of water plants on every continent but Antarctica. When it comes to the global water crisis, he sees things that the rest of us don't. Wetterau is a quietly charismatic vice president at CDM Smith, the global engineering giant, but he's a water warrior too. He is driven, professionally and personally, to help the world confront the water crisis head-on.

On a recent sunny California afternoon, he and I embarked on a wide-ranging conversation about the idea that we now have the technology to turn wastewater from showers, sinks—and, yes, even toilets—into drinking water. Our discussion quickly zeroed in on a California law that required perfectly drinkable recycled sewage to be blended with groundwater before it could be processed into drinking water. Never mind, Wetterau said, that the purified wastewater is often *cleaner* than the groundwater with which it is being mixed. To Wetterau there was something ironic about pumping perfectly drinkable recycled water underground just because people didn't trust it.

"Do you drink beer?" he asked suddenly.

"Sure," I said.

He whipped out a bottle from his backpack.

"FAT Californian," the label read. "American Oatmeal Porter." The beer was brewed by Wetterau's company—from purified sewage.

I was intrigued.

The techy lingo on the bottle read "California's full advanced treatment process has become the gold standard in potable reuse, converting wastewater flows into safe drinking water supplies for the water-stressed state."

The label boasted that the beer was brewed with water that was "cleaner than the most pristine drinking water supplies." Wetterau told me that the porter was made from recycled wastewater that had *not* been mixed with groundwater. The purified sewage went straight from a treatment facility into the beer, without being blended in a so-called environmental buffer like an aquifer. That sounded like toilet to tap, I thought, a phrase that some have used to disparage sewage recycling. While I've always found the expression to be blunt and descriptive, the water reuse community reviles the term as inaccurate, pejorative, and simplistic. Wetterau said that California regulators gave his company special permission to brew the beer with the 100 percent unblended purified sewage as long as CDM Smith didn't sell it. It was fine to give it away to people like me to help make the case that so-called direct potable recycled water was safe to drink.

"Would you drink it," he asked?

"Of course."

That evening I snagged Wetterau's beer from my hotel minifridge and poured a tall frothy glass of brown porter. Licking the tan foam

Figure P-1. CDM Smith's American Oatmeal Porter, brewed with purified sewage. (Photo by Peter Annin)

from my lips, I couldn't help wondering how many bladders and bowels some of those water molecules had passed through before gracing my gullet. That mental image was easier to concoct given the brown color of the beverage. *But the beer tasted great.* I couldn't tell it was brewed with recycled water.

So yes, I did drink it, and given the rapid growth in the nation's water recycling movement, increasingly the question will be whether millions of Americans are willing to do so as well.

CHAPTER 1
Dead Pool

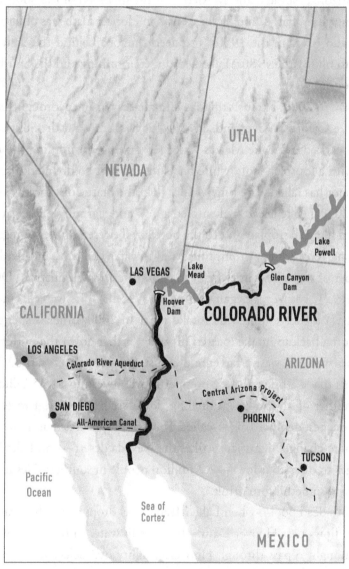

The Colorado River

PERCHED JUST OUTSIDE LAS VEGAS, Lake Mead is the largest reservoir in the United States. Sprawling for more than one hundred miles behind Hoover Dam, it stores water for twenty-five million people in California, Arizona, Nevada, and Mexico. It is the lifeblood of the Southwest's thriving cities, farms, and factories. When Hoover Dam was built on the Colorado River in the 1930s, the deep canyons behind the dam took years to fill. By the 1980s, Lake Mead was brimming and the Southwest was booming.

Then, in 2000, a transformative climate-driven "megadrought" swept over the Colorado River and stayed—for decades.[1] But the Southwest's boom carried on. Lake Mead's water level plunged, rimming the reservoir with a fourteen-story white bathtub ring that boldly marks how far the water has fallen. The ring is taller than the Statue of Liberty.[2] By the early 2020s, Lake Mead had entered a new and alarming phase. In what seemed like a dystopian race to the bottom, the reservoir began regularly breaking record lows, prompting painful water cutbacks and sending a shudder of water insecurity from the Southwest to Washington, DC.

In 1971, Las Vegas began pulling 90 percent of its water from Lake Mead via a twelve-foot-wide intake pipe. In 2000, officials spent millions on a backup intake located fifty feet farther down the underwater canyon wall.[3] It wasn't far enough. Just five years later, rattled Vegas officials made a bold climate-prepper move: investing $1.35 billion on the mother of all intakes, tunneling down six hundred feet to tap the lake's deepest depths. This so-called Third Straw ensures that if, or when, a climate-change nightmare strikes the Nevada desert—and Lake Mead plunges to disastrous lows—2.2 million people in southern Nevada will still have something to drink.[4]

That's dead pool—when Lake Mead's level drops so far that water no longer flows past Hoover Dam.[5] There's still water in the reservoir, just not enough to pass through Hoover's mighty hydroelectric turbines or even slip by the dam. That would be a water catastrophe. But if Hoover's

Figure 1-1. By the early 2020s Lake Mead had reached a new and alarming phase, repeatedly breaking low-water records. (Photo by Peter Annin)

turbines fall silent, an electricity emergency would cascade from the Rockies to the Pacific. The dam provides power to 1.3 million people in Nevada, Arizona, and California.[6] If the river stops flowing past the dam, millions of acres of farmland in California and Arizona would dry up, cratering produce deliveries throughout North America and dramatically increasing water tensions with Mexico. Much of the United States' vegetables are produced by Lake Mead's water, especially in the winter. So, if dead pool ever arrives at Hoover Dam, it will be a triple-whammy—water, power, and food emergency—the likes of which the United States has never seen.

For decades, dead pool seemed like a far-fetched notion. Then came the summer of 2021. Not only did Lake Mead break a record low in June of that year, but in July, Lake Powell—the equally massive Colorado River reservoir farther upstream—also broke an all-time low.

Never before had the nation's two largest reservoirs been so low at the same time. Suddenly, these enormous artificial lakes—which together provide a crucial water supply to forty million people—seemed shockingly vulnerable. Things grew even worse in 2022. In a sobering historical moment, in April of that year, Lake Mead fell so far that the original water intake pipe—that Las Vegas installed back in 1971—began poking awkwardly out of the water.[7] The following month, a body was found in a barrel that had emerged from the receding reservoir—captivating the nation.[8] A gun turned up too.[9] Officials said it looked like a 1970s mob hit.[10] As the Colorado River struggled through the worst drought in twelve hundred years[11] and water officials scrambled to respond, Lake Mead began revealing some of its darkest secrets, making the situation seem even more surreal.

John Entsminger is one of several key southwestern officials who worry about the river. As general manager of the Southern Nevada Water Authority, he is responsible for providing water to the 2.2 million people in the Las Vegas metro area. He told me that he is "extremely concerned" that dead pool may soon be upon us. It could happen in two years, or ten, but his tone made it seem more like a "when" than an "if." We both agreed that most people don't realize just how tough things have become on the river. Given that, I asked him to describe what dead pool would look like. "If Mead gets to dead pool, that means twenty-five million Americans downstream of Hoover Dam lose access to the Colorado River. The states of Arizona and California and the country of Mexico have no access.... There's no law that can be passed, there's no political speech that can be given, that will change the laws of physics when you can't pass water through Hoover Dam—that's what's at stake."

The long-term scientific projections for precipitation and runoff in the Colorado River watershed are depressing. In 2012 the US Bureau of Reclamation estimated that by 2060 the river could face an *annual* shortfall of a trillion gallons,[12] or roughly half of what Arizona uses per

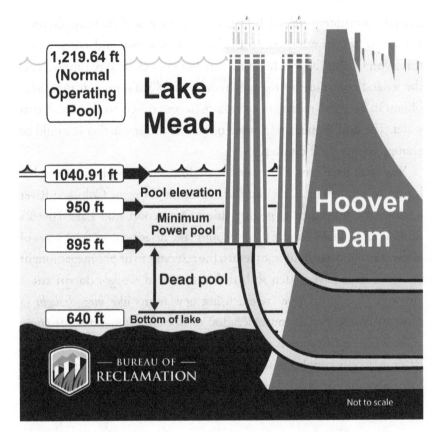

Figure 1-2. If the level of Lake Mead falls below minimum power pool, the mighty turbines ensconced in Hoover Dam will fall silent, impacting the electrical grid from the Rockies to the Pacific. Dead pool occurs when the level of the reservoir falls so far that water no longer flows past the dam.

year.[13] That prediction was for a river that is already so overtapped that it regularly no longer flows to the Sea of Cortez.[14] But in the climate change era, sometimes the bad news comes early. In 2022 the river's situation had become so dire that the federal government told the seven Colorado River states to come up with a plan to conserve an additional 652 billion to 1.3 trillion gallons per year—*four decades sooner* than the 2012 report anticipated.[15] When the states failed to act, in 2023 the federal government helped broker a temporary deal creating almost a trillion gallons in water cuts for Arizona, California, and Nevada over three years.[16] But those reductions were not even close to the annual cuts the federal government originally requested. What's more, it took $1.2 billion in taxpayer money to coax farmers and others to relinquish that water. The deal immediately raised questions about whether it would be enough to "stave off disaster."[17]

Yes, there have been a few wet years—like the atmospheric rivers that hit the Southwest in 2023. But the overall trend for the Colorado River is clear: the watershed is getting drier. One report said Lake Powell's water level had fallen so far that it would take fifteen consecutive years of above-average precipitation to return the reservoir to its pre-megadrought level.[18] Things have gotten so bad that the word *drought* doesn't cut it anymore. Scientists have started using new terms like *megadrought* (a drought lasting two or more decades) or *aridification*—droughts come and go, but aridification is a new normal. In coming years, the Colorado River may only supply enough water to maintain levels in just one massive reservoir, not two.[19] Dead pool could become normal, threatening long-term capital investment throughout the Southwest.

⌐

Nevada is running out of water options. Arizona and California are too. In the face of a relentlessly changing climate, all three states in the Lower Colorado River Basin are scrambling to diversify their water portfolios. They are particularly concerned about the reliability of the Colorado as a primary source of supply. What other water options do they have? The era of environmentally disastrous long-range, large-scale water diversions is over—or at least it should be—so piping in rescue water from elsewhere isn't an option. Arizona has stored more than a trillion gallons of backup water underground.[20] Nevada has banned grass in many areas.[21] California has built twelve desalination plants[22]—others are on the way[23]—but none of that will be enough. All three states have banked years of extra water in Lake Mead, but that doesn't help California and Arizona if dead pool hits because then their backup water can't get past the dam. The long-standing practice of buying heirloom water rights from farmers remains an option, but most farmers with the best rights don't want to sell. Nevertheless, pressure for more ag-to-urban water transfers is ever present, especially if the long-term viability of rural communities can be protected. Yes, huge strides can still be made with water conservation—especially in the agriculture sector—but the Southwest's water situation is so dire that full-throttled conservation alone won't replace the loss of a trillion gallons per year.

Something else must be done.

Enter purified sewage. Thanks to climate change, never before has something so foul looked so good. Wastewater has become one of the most popular new water options in the Southwest's scramble for additional supplies. Nevada, California, and Arizona cumulatively are investing billions in water recycling. Las Vegas currently recycles 99 percent of water used indoors. Los Angeles has pledged to recycle 100 percent of its effluent by 2035. Scottsdale, Arizona, has built a cutting-edge potable water recycling plant. The water crisis has become so acute that in many parts of the United States there are only two realistic options left for new water supplies: the ocean and the toilet.

Ocean desalination is an important option, but it is often more expensive than water recycling and has more negative environmental impacts. Those impacts include a much larger carbon footprint and a more heavily concentrated brine by-product that is dumped back into the sea. There are also concerns about fish mortality due to the entrainment that occurs when juvenile and larval fish get sucked up through screens that cover the opening of a desalination plant's large seawater intake pipes. Water recycling produces a brine by-product too, but much less so. (Only about 15 percent of recycled water ends up as brine, whereas a desalination plant discharges at least a gallon of brine for every gallon of freshwater produced.) And since no ocean intake pipes are involved with water recycling, fish entrainment is not an issue.

Water recycling is usually cheaper than ocean desalination, and its cost has become increasingly competitive with imported water from places like the Colorado River. But unlike the Colorado River, recycled water is reliable—there will always be sewage—which is why officials increasingly see any additional cost as being worth the added security of a drought-resistant supply. What's more, recycled water is available to everyone, including landlocked cities that are hundreds of miles from the ocean, where seawater desalination is not an option. "Unless we move into a new era of thinking about water, we're not going to solve our water problems," warned Peter Gleick, senior fellow and cofounder of the Pacific Institute and one of the nation's leading water experts. "We can no longer ignore these nontraditional solutions like water reuse."

The nation's water recycling boom stretches far beyond the Southwest. Texas hosts an arc of water tension that starts in El Paso—home to seven hundred thousand people—and stretches through the West Texas oil patch, brushes over the panhandle's depleted Ogallala Aquifer, and ends at Wichita Falls, near the Oklahoma border. Water recycling is thriving throughout this parched region.

In the Southeast, Florida has designated almost 70 percent of its land-mass as a Water Resource Caution Area,[24] highlighting those portions of the state that already have water shortage problems or are expected to in coming years. State projections show that officials will need to find another billion gallons of water, *per day*, by 2040.[25] The state has long used recycled water on golf courses and lawns, but it also overhauled statutes in the early 2020s to make it easier for cities like Jacksonville and Tampa to produce drinking water from sewage.[26]

Farther up the eastern seaboard, Virginia has long been a quiet leader in water recycling. That is particularly true in parts of suburban

Washington, DC, which have depended on recycled water for decades. But now, in southeast Virginia, officials plan to pump one hundred million gallons of recycled water underground daily to replenish the troubled Potomac Aquifer,[27] one of the most important drinking water sources on the East Coast.

These efforts in Florida and Virginia show that the water-reuse boom is not just limited to "dry" states. It's happening in "wet" states too. "I think eventually, in America, water recycling will be considered normal everywhere," predicted Greg Wetterau of CDM Smith, the global engineering firm. "It's happening in California, Texas, and Florida—anywhere that is water-stressed. But we're already seeing it in Wisconsin and Ohio, so it's not just the places that you might expect," he told me. Although some communities have depended on recycled water for decades, the trend is going national quickly, reaching a long-anticipated moment in water management history where reuse is seen as a much more sustainable water supply option than the environmentally damaging long-range water diversions that have been a hallmark of the American West for more than a century. Robert Glennon, a University of Arizona water expert, said that the recycling boom "signals an end to the era of addressing water shortages by importing water from far-flung places and initiates a long-anticipated era of reusing locally available supplies."[28]

Water reuse is expanding globally as well. Singapore has been an international leader in water recycling for years. Namibia, Australia, and South Africa have too. In Israel more than 85 percent of the nation's water is reused, most of it going to agricultural irrigation.[29] In 2020 the European Parliament endorsed a major expansion of water reuse, potentially more than quintupling the amount of recycled water on the continent—to 1.7 trillion gallons per year.[30] "Israel is definitely the leader in non-potable reuse," Wetterau told me, but when it comes to drinkable recycled water, "globally, it's the US who is the leader—by far." California leads the United States in potable water recycling, and

Orange County leads California, making it a global leader in potable water reuse.

Even though water reuse technology has been around for years, it has long struggled with public acceptance. But the combination of extreme drought, climate anxiety, major investments in public relations, and technological advances have softened opposition to water recycling in key states. The treatment process in Orange County, for example, is extensive, and multitiered. The water is first pushed through high-tech filters that screen out microscopic protozoa, bacteria, and viruses and then it is purified with reverse osmosis, which removes any remaining viruses. Also removed at this step are any "forever chemicals," such as per- and polyfluorinated substances, or PFAS. In the next step, the water—which is already purified—is treated with hydrogen peroxide and zapped with ultraviolet light to purify it even further. The ultimate product is akin to distilled water, so pure, in fact, that elements need to be added back into the water so that it doesn't leach minerals from municipal plumbing systems on the way to people's homes. As climate change continues to disrupt the national water balance, Rabia Chaudhry, a water reuse expert at the Environmental Protection Agency, predicts that "we should be anticipating that climate-intensified events will shape public opinion about water reuse." She said, "Water reuse will increasingly no longer be optional."[31]

It will become necessary. For millions.

So here we sit at a time when climate change has transitioned from a complicated and much-debated scientific theory to a sobering, bald-faced reality that threatens the national water supply. Sewage is coming to the rescue. Throughout human history we have worked tirelessly to distance ourselves from wastewater, and for good reason: it's dangerous and disgusting. But thanks to climate-driven water scarcity—and impressive advances in treatment technology—purified effluent has emerged as a leading weapon in the war against water scarcity. Recycling can't solve all the world's water woes. But in the United States—especially in the Sun Belt—it is seen as the right technology, at the right time, to give officials water-planning breathing room as they contemplate what the next fifty years of climate change might bring.

In short, wastewater is just too precious to waste anymore.

CHAPTER 2

"Gulp!"

This editorial cartoon shows the kind of negative publicity that emerged from San Diego's "toilet-to-tap" controversy of the 1990s. (By permission of Steve Kelley and Creators Syndicate, Inc.)

THE HEADLINE SAID IT ALL: "Water from (gulp!) where? City aims to make sewage drinkable."

The date was July 6, 1997. San Diego was in the midst of a years-long, $152 million trailblazing effort to add purified sewage to its drinking water. "Wastewater officials say 5,000 hours of experiments just completed prove conclusively that sewage flushed down toilets and drains can be made safe to drink," the article said. "The city hopes such highly treated wastewater will begin quenching San Diego's thirst by the end of 2001 in what would be the first such sewage-water recycling program in California."[1] Under the ambitious plan, the city would pump millions of gallons of purified sewage into a reservoir, where it would mix with water piped in from the Colorado River and Northern California. Then this blended water would be treated again and delivered to residents via the city's drinking water pipes. Officials called it "water repurification." Critics called it "toilet to tap."

It was a pioneering proposal for the late 1990s. How would the public react? The *San Diego Union-Tribune* story included supportive quotes from city officials, the local medical society, the Sierra Club, even a federal judge. "There's really no question, from a technological standpoint, that we can build a system that will provide safe, reliable and cost-effective water from wastewater," assured Paul Gagliardo, the project's manager.[2] But the *Union-Tribune's* most sensational quotes came from Bruce Henderson, a former San Diego councilman. Henderson had become one of the city's most polarizing figures,[3] and he now emerged as a leading critic of the groundbreaking water recycling plan. He lambasted the project as a "Dr. Frankenstein" experiment and accused city officials of turning San Diegans into unwitting "guinea pigs" who were "being forced to participate in a health experiment" without their consent.[4]

Pretty alarming stuff. Numerous articles had been written about San Diego's water recycling program before July 1997 with little incident or opposition. Many more were written after. But in reviewing hundreds of articles about San Diego's decades-long water woes, it's clear that Kathryn Balint's "Gulp" article—and a "toilet-to-tap" graphic that appeared with it—marked a clear turning point. That's when San Diego officials began losing control of the narrative over their project. That's the problem with water recycling. When you are trying to convince people that it is perfectly safe to turn sewage into drinking water, it is very easy for opponents to hijack the narrative. In subsequent months, the bad publicity would build and build, to the point where it eventually became politically unsustainable for officials to support the project. Then it died. Decades after San Diego's seminal experiment in water recycling, the project's implosion still stands out as the leading example of how quickly public perception can spiral out of control with a well-meaning and progressive water recycling program.

San Diego is known for its thriving waterfront and pleasant year-round climate. During the last century as it became the home port to the US Navy's Pacific Fleet,[5] the city attracted a high concentration of military installations and defense contractors,[6] while the temperate weather supported robust agriculture[7] and tourism industries.[8] Those attributes attracted so many people since World War II that San Diego has ballooned into California's second largest city, topping out at 1.4 million today.[9] The growth was fueled by water.

But from a water standpoint, the pleasant climate had its drawbacks. City officials estimated that San Diego's ten inches of annual rainfall was enough to support a population of just one hundred thousand people.[10] The last time local supplies met regional demand was 1946. So it was no surprise that by the 1990s, cursed as it was with paltry surface and groundwater resources, San Diego was piping in more than 85 percent of its water from hundreds of miles away.[11] Most of that imported water came from the Colorado River, and the rest came from Northern California via the sprawling State Water Project, along with some nominal local supplies.[12] Sitting at the southern extreme of the state, San Diego was literally at the end of the pipe.

As far back as the 1980s, water reuse was viewed by civic leaders as a promising, local, drought-resistant, water supply option.[13] That's when officials started building a series of water recycling pilot plants to use nonpotable recycled water for irrigation.[14] By the mid-1990s San Diego officials had decided to build two full-scale plants that could produce up to forty-five million gallons of nonpotable recycled water per day[15]—in part because they had to: the federal government was requiring it as part of a complicated Clean Water Act sewage discharge settlement with the US Environmental Protection Agency (EPA) and local environmental groups.[16]

But the nonpotable water recycling program soon bogged down. Recycled irrigation water needs to be distributed through a completely separate purple pipe plumbing system to prevent it from mixing with drinking water. Because this recycled water is nonpotable, it is intentionally delivered in purple pipes to make it clear to everyone that the water is not fit for drinking. Laying purple pipe in urban areas is expensive, and San Diego did not install enough pipe to get rid of all its recycled water.[17] Instead, controversially, it dumped that valuable irrigation water into the ocean. After spending millions on purple pipe and barely increasing the use of recycled water, city officials made a historic decision. Rather than spend more money expanding an entirely new urban plumbing system for nonpotable irrigation water, they chose to invest that money on advanced water treatment technology instead.[18] The idea was to purify the sewage to the point where people could drink it. That water wouldn't require a new set of expensive extra pipes. It could be delivered to homes through the existing drinking water system instead.

As San Diego moved closer to launching its potable water recycling program, a fascinating cast of characters emerged to dominate the headlines for the next several years. On one side, you had the City of San Diego's water recycling team, led by Dave Schlesinger, a highly capable can-do retired Navy veteran who survived three tours of duty in Vietnam. Trained as a civil engineer, he was also an experienced supervisor. By the end of his naval career, he was leading a team of a thousand people. As Schlesinger prepared to segue out of the military, he was recruited to take over San Diego's Metropolitan Wastewater Department at the most complicated time in its history. Not only was he juggling the litigation with environmental groups and the EPA, but he was also trying to build an immense potable water recycling program. "What intrigued me," he told me decades later, "is that it was clearly going to be the largest public works job in the city's history." Schlesinger's role was to keep the big picture in mind, fend off the lawyers, brief the mayor and other politicos, and deal with the media, all while keeping money flowing to the project. There were moments when it seemed like he was spending as much time with attorneys as engineers, but he loved the challenge of the work. "Next to the Navy," he said, "that was the best job I ever had."

At least at first.

Schlesinger's go-to guy was Paul Gagliardo. Tall and outgoing, Gagliardo was a Long Island transplant who moved to California in the late 1970s armed with a mechanical engineering degree and a fistful of enthusiasm. But he had to adjust to the dry climate. "I moved down in April and went back to visit my folks at Christmas," he told me, "and on the plane home, I realized it had never rained once." Gagliardo's primary task was to build the water purification treatment process, using microfiltration, reverse osmosis, ozone, and peroxide treatment, followed by chlorine for added purification. He was also responsible for convincing state health officials—and ultimately the public—that the city's purified sewage would be safe to drink. As a midcareer engineer in the Wastewater Department, he threw himself into the historic effort. "It was great," he told me. "It was some of the most fun I ever had."

At least at first.

On the other side was a colorful cast of accomplished headline-grabbing critics, starting with Henderson, the former city councilman. A Berkeley Law grad, Henderson would become such a polarizing figure in San Diego that one local paper would suggest that he be locked up in Guantanamo Bay.[19] His sensational criticisms of the city's emergent water recycling program would mark one of many high-profile disputes that helped brand Henderson as the city's most notable naysayer, although he preferred the to be called a whistleblower. As his own attorney once put it, "Bruce Henderson has been vilified more than any other person I've ever seen in San Diego politics."[20] When I interviewed him for this book, Henderson heartily agreed that he remained a polarizing figure years after leaving politics. His role in killing the city's water recycling program helped solidify that reputation. "You always have to remember the general rule in engineering: if it can go wrong, it kind of will," he told me.

Then there was Stephen Peace, a high-profile Democratic state senator. Peace's early career included the production of B movies—most notably the *Attack of the Killer Tomatoes!* series, which also featured *Return of the Killer Tomatoes!*, *Killer Tomatoes Strike Back!*, and *Killer Tomatoes Eat France!* In a classic California moment, his moviemaking career transitioned to the state legislature. A San Diego native, Peace was known as a skilled debater, and today he is remembered as the most intimidating figure to oppose the city's water recycling program. "These guys, however well-intentioned," Peace told me years later, "hadn't laid the groundwork for this.... It couldn't be survived, politically."

⌣

But in the early 1990s, as the "water repurification" program was still getting started, it percolated along with little resistance. "We're not getting any public outcry," Schlesinger reported.[21] As Gagliardo worked at perfecting the technology, Schlesinger openly wondered if public perception might change. "I think the biggest obstacle will be some group that, for whatever reason, ... says, 'This is crazy. They're going to have us drinking our sewage."[22] Schlesinger and Gagliardo never doubted the purified water's safety. "I drank it all the time," Gagliardo told me. "I let my kids drink it."

Early on, the biggest challenges were internal, especially in terms of coordinating the work of Schlesinger's wastewater team with the Water Department. "We were producing water that was higher in quality than the Water Department was producing, but we couldn't say that," Gagliardo told me, "because then we'd be dissing the Water Department.... So, we had to be very careful about how we presented the product we were generating," he said. "We had to use weaselly words. We couldn't go out and say, 'This is like distilled water.'" Which it was. It didn't help that the Water Department felt it was superior to the Wastewater Department—even before the sewage purification project was launched. "There's a hierarchy in the utility sector," Gagliardo explained. "Trash is on the bottom, wastewater is one step up, and then water is one step up from there.... They really didn't want to have anything to do with what we were doing."

In 1997 Schlesinger and Gagliardo released a definitive report.[23] The document made it clear—to the public and the press—that the project was moving from the theoretical phase to a reality. That's when the media coverage began to change, starting with the *San Diego Union-Tribune's* "Gulp" article in July of that year. From then on, just about every article about the project mentioned the term *toilet to tap*, and opposition increased. "The problem here isn't technology. It's terminology," wrote one columnist. "No one will drink 'repurified sewer water.'"[24]

News from outside San Diego didn't help. In 1993 Milwaukee was hit with the largest waterborne disease outbreak in US history when

Cryptosporidium swept through the drinking water system, sickening hundreds of thousands and killing sixty-nine more.[25] The Milwaukee crisis set back public confidence in *traditional* urban drinking water systems and made it even more difficult to pitch new-fangled programs like San Diego's.

What's more, newspapers across the United States at that time were filled with articles about "endocrine disruptors" and trace pharmaceuticals that were detected downstream of sewage treatment plants, where fish were being discovered with abnormal sex organs. In the face of these concerns, San Diego officials assured people that all those contaminants would be removed by the city's water purification process. More importantly, the recycled water would be much cleaner than the raw water shipped in from the Colorado River with which it was to be mixed. The Colorado contained effluent from hundreds of sewage treatment plants, including those for Las Vegas. In fact, a popular line in San Diego at the time was "What happens in Vegas *doesn't* stay in Vegas." Instead, it flows downstream for San Diegans to drink. But increasingly the city's reassurances failed to resonate with average citizens. Public opposition mounted.

⌣

Then politics, equity, and class entered the mix. In December 1997, Peace worked with his state Democratic colleague Howard Wayne to organize a legislative hearing in San Diego about the project. In the curtain-raiser articles announcing the hearing, Wayne echoed earlier comments from Henderson, saying that he too wanted "to make sure we're not going to be used as guinea pigs." Peace chimed in by calling the recycling plan "cockamamie." "This is a situation," Peace complained, "where the bureaucracy got well ahead of the policy makers."[26]

At the emotional public hearing, the discord ramped up further, with an environmental justice twist. Peace and Henderson alleged that because of the graduated way officials planned to roll out the purified sewage program, the recycled water would be sent to low-income neighborhoods first before eventually being expanded to wealthier neighborhoods later. "It exacerbates the division in our society between those who are affluent and those who are not," complained Henderson.[27] Unfortunately for Schlesinger and Gagliardo, that allegation fed straight into the "guinea pig" narrative. "The south part of town was mainly African American," Gagliardo said. "That became the bigger issue from the public's perspective.... 'Us poor folk in the south part of town are getting the rich folks' sewage.'" Gagliardo and Schlesinger said that they weren't intentionally targeting these communities of color; it was just that, given how San Diego's plumbing was laid out, it made sense, logistically, for those neighborhoods to be brought online before the others. "It was just a plumbing issue, pure and simple," Gagliardo told me. "But that wasn't the way it was perceived."

That charged theme dominated the hearing all evening. One local resident stood up in front of 250 people with bottled water in her hand. "The rich will be drinking this," she scolded. "The rest of us will be participating in a long-term medical study." Tensions increased as Peace took the microphone. "If you think you can get away with taking the effluent of the rich and selling it to the poor as drinking water, it's dead on arrival," he threatened. Gagliardo remembers a long line of Black ministers laying down scathing and articulate public comments denouncing the city's plans. "I mean the rhetoric was fantastic—it was heartfelt—but it was such a show," he said. The meeting went on for hours and hours. "One after another, they would come up, and they were invoking the Tuskegee syphilis experiments," Gagliardo remembered. "I mean, it was really quite extraordinary."

As the project's primary spokesperson, Schlesinger took the brunt of the criticism. Despite the heated emotions in the room, he attempted to rebut the barbs methodically and factually. But the facts were steamrolled by emotion. "[He] was just getting *murdered* up there!" Earle Hartling, a Los Angeles area water official, told me. The hearing received so much advance publicity that Hartling had traveled down from Los Angeles to see how things might transpire. "[Schlesinger] was trying to give rational, engineering, scientific answers that nobody wanted to hear," Hartling remembers. "Everybody was nuts!"

A local Sierra Club official tried to support Schlesinger, saying that San Diego was running out of water options. "There is no more water," she warned. "We have to come up with local solutions." One city councilwoman pointed out that San Diegans were already drinking "the effluent of the affluent," as well as the effluent of the "poor and the sick"—all of which was being discharged into the Colorado River, San Diego's primary water source. Peace fired back that her claim was "worthless at best, and extraordinarily deceptive at worst" because the percentage of purified wastewater would be much higher in the local reservoir than it would be in the Colorado River.[28]

Headline writers had a field day. "In some neighborhoods, you'll get the effluent of the affluent," blared the *Union-Tribune*.[29] In the wake of the hearing, a new water recycling opposition group emerged: the "Revolting Grandmas." These passionate women quickly became some of San Diego's more memorable water recycling opponents. "Why the hell do we have to drink our own sewage?" one asked.[30]

And so it went. San Diego's water recycling plan was mired in front-page controversy and blanketing the evening news. Local columnists who supported water recycling accused former councilman Henderson of "alarmist claptrap"[31] and called Peace a "political opportunist."[32] But those columns had little effect. The narrative had shifted, seismically, and the water reuse program was now in peril. "We could see the handwriting on the wall," Gagliardo remembers. "We walked out of that meeting saying, 'It's dead—dead in the water.'"

As it happened, 1998 was a local election year, and faced with the rising controversy, officials who favored water recycling softened their support. They stopped pushing so hard for the project and asked for more studies on the idea instead. Sensing a political opportunity, Henderson announced that he was running for city council again—challenging an incumbent water recycling supporter. City officials decided to delay the release of a much-anticipated environmental impact report, fostering criticism from Henderson and others that the city was trying to keep water recycling out of the headlines during the election season.

Then, in a highly damaging turn of events, the prestigious National Research Council released a much-anticipated 1998 report on water recycling. The council declared that indirect potable reuse—the technical term for water recycling projects like San Diego's—should be considered "an option of last resort."[33] That was far from the resounding endorsement that San Diego's embattled water recycling team needed. "It is important to recognize that although indirect potable reuse can be considered a viable option," the council said, "many uncertainties are associated with assessing the potential health risks of drinking reclaimed water."[34]

"Many uncertainties"? That highly cautionary language contradicted the water recycling reassurances that the public had been hearing from Gagliardo and Schlesinger. The council's authoritative report suggested to a cautious general public that the technology was not ready. During the months that followed, San Diego's water recycling program was battered in the headlines, even though one of its leading critics, Henderson, lost his reelection bid.

In a last-ditch move to head off the criticism, Schlesinger announced that the city had changed its mind: the purified sewage would *not* be delivered gradually, a few neighborhoods at a time, but rather it would be delivered to the entire city at once. "This will eliminate their concern that they're going to be guinea pigs," Schlesinger said. The move would delay the launch of the water recycling project by a few years, but officials hoped it would quell the stinging complaints that the purified effluent would go to communities of color first.[35]

It was not enough. In December 1998, Republican mayor Susan Golding put the entire project on hold. "I don't think we're anywhere near satisfying ourselves that it's the right thing to do," she said. "*I don't feel satisfied that it's the right thing to do.*"[36] Her announcement came almost exactly a year after the emotional public hearing. In the end, twelve months of criticism from Peace, Henderson, and others created such a media sideshow that public confidence in the project collapsed. "The city of San Diego has spent more than $15 million, has accumulated twenty years of scientific research and has waged a major public relations campaign to convince politicians as well as the general populace that sewage can be made safe to drink," the *Union-Tribune* reported. "But all that might be going down the drain."[37]

In San Diego, water recycling had become too toxic for elected officials to support. "Toilet to tap is hazardous to the political health of anyone who touches it," warned one consultant. "It is a politically—not scientifically—but a politically indefensible idea."[38] George Stevens, the Black city councilman whose district was on the south side of town, called the plan "an ill-conceived idea whose time has not yet come."[39] Peace simply declared, "Bad projects eventually kill themselves."[40] Subsequent San Diego water recycling stories read more like obituaries than news articles. In early 1999, in the wake of the mayor's plug-pulling, the formerly bullish city council cut off funding for the water recycling project.

Finally, in May of that year, the city council voted to kill San Diego's once-promising water reuse plan. "Dead, killed, buried, gone with no chance of coming back to life—ever, forget it," read one over-the-top article about the vote. "That's what San Diego City Council members said yesterday of a reviled and publicly distasteful proposal to turn sewage water into drinking water."[41] During that meeting, council members were so bent on terminating the water recycling program that they pushed city staff to confirm that they were not somehow secretly keeping the project on life support behind the scenes. "We have killed that project," assured the deputy city manager. "We will not be spending any money on it."[42]

It was a shattering blow to Schlesinger and Gagliardo. "I'll be honest with you," Schlesinger told me, "I was very, very disappointed.... Those were tough times." He was particularly worried about how Gagliardo was dealing with the news. "I was devastated—absolutely devastated," Gagliardo told me during a kitchen-table interview more than twenty years later. "You put so much of your energy into it—you lived it for so long—that you're like, 'Really?'" For Gagliardo, who had full faith in the purified water—even allowing his children to drink it—there was something personally insulting about his water being equated with the horrific Tuskegee syphilis experiments. "It's been a long time since I talked about this," he confessed. "You asked me how I felt? I think you can tell by my attitude—I'm still pissed off."

⤳

What lessons can be learned from the collapse of San Diego's groundbreaking water recycling project? Beyond the emotional debate recapped above, other forces helped eliminate the plan. The project was born in a severe drought, but interest waned as the drought subsided and San Diegans felt more secure about their far-flung water supply. Because the city's water recycling controversy peaked during an election year, the plan became even more vulnerable as campaigning politicians grew squeamish under withering criticism. In particular, the release of the National Research Council's hesitant water reuse report—in the middle of an election—made for a very difficult confluence of events for Schlesinger and Gagliardo to overcome.

Despite San Diego's extensive investment in public relations, a handful of articulate, influential, and media-savvy critics were able to shatter public confidence in the water recycling plan. Yes, the city made some strategic mistakes, but in the end, the San Diego water reuse story is a classic case of well-meaning, reasoned, science-based messaging that loses out to brilliantly simplistic and sensational criticism. Today the San Diego case study is the most compelling example of how important it is for water recycling programs to back up their top-notch engineering technology with an equally sophisticated communication strategy. San Diego's case is well known in the water world, and it set back the city's water recycling program for two decades. "In hindsight there needed to be more outreach," said Peter MacLaggan, former head of the California WateReuse Association and a longtime player in San Diego's water circles. A few "very innovative thinkers" realized that some local residents weren't on board, "and it was easy to mobilize them and cause fear and anxiety, and that's what the setback was all about."

Most of the San Diego takeaways focused on communications and how city officials lost the narrative over their cutting-edge project. "I think the lesson that came out of the 1990s was that maybe we overestimated the strength of our hand," MacLaggan added. "Things like this, that sort of change the way you think about the world, take time for people to get comfortable with." To Gagliardo, the communication strategy was hampered by too much defense. "I think you *have* to go on offense," he told me, "because otherwise somebody will come out of the woodwork and come up with something that will catch on fire, and then you can't stop it. The toilet-to-tap thing—that line became the de facto moniker for the project." What's more, Gagliardo said, the city should have taken its messaging straight to the people, including those on the south side of town. "We spent a lot of time preaching to the choir," he told me. "We never did the work to go and actively seek out the people who did not agree with us.... The more they feel like they're ignored, the worse it is."

Schlesinger had a different take. He thinks the city put too much emphasis on water recycling when it would have been more effective to couch it as one addition to a portfolio of different potential supplies, including desalination and reservoir expansion. "I think we oversold the issue of recycled water being the salvation of an area like San Diego that has to import all its water," he told me. "I think we made a mistake of trying to sell recycled water as a panacea."

Sara Katz was the lead communication consultant hired by the city to help. Today she is one of the leading outreach experts in the industry, but she admits that San Diego spiraled out of everyone's control. In the early years she worked hard with Schlesinger to build a broad coalition of supporters. "We checked off every box that you would in a political campaign," she said. But the city had a big problem with the *Union-Tribune*. One editorial page writer was a strident water reuse opponent. No matter how much the city met with him, Katz said, the "yuck factor" would constantly be a theme in his editorials. Complicating matters, many city council members were reluctant to lead on the issue. Katz said that councilors would tell her, "I get it but, you know, it's yucky. I want somebody else to be the voice. I want somebody else to take the leadership role."

Katz takes a cynical view of the now-infamous public hearing that served as the beginning of the end. She told me that the driving force behind the hearing was politics, not a genuine concern about public health. Katz and Schlesinger maintain that Peace and others were trying to use the toilet-to-tap controversy to taint Mayor Golding's administration. "Steve [Peace] was the instigator of all this," Katz told me. The hearing, which she described as "eviscerating," was designed to "make Susan Golding look bad.... Pure politics." She said that once the narrative had shifted, there was only so much that a consultant like her could do. That was complicated by Mayor Golding never personally pushing the water reuse project, leaving it up to Schlesinger to promote it instead. When you combine that with an equally reluctant city council, any communication strategy is going to be handicapped. "We had built a very strong coalition," Katz said. "We had done a great job with visuals, with tours, empowering those voices that had followings to help us tell that story, but you know, at the end of the day, it's politics."

I reached out to Peace to see what he thought. I met the former state senator at a waterfront restaurant in San Diego on a rare rainy day. When I told him that numerous people accused him and other critics of leveraging the hearing for political gain, he denied it. Think about it, he said—the hearing took place in December of a nonelection year. He said that if it was meant to be political, the hearing would have been scheduled during the election season. "You wouldn't do a hearing for just pure, raw, nonmeritorious political reasons," he said. Rather, he told me that the city's problem was simple: officials had not laid enough groundwork with the public for their water recycling program. "The worst thing you could do with a good idea is kill it with your personal enthusiasm by not understanding that.... the public has to embrace what you want to do," he said. Then, toward the end of our conversation, unsolicited, he offered up this final thought: "But really, how pathetically awful are you if you have the benefit of truth and you can't sell it?"

Henderson was equally defiant. He and I also met at a restaurant in San Diego. To Henderson, the program was too expensive, conservation would have been cheaper, and it was unfair and politically unsophisticated to send the water to communities of color first. "Conservation is the best way to get the goddamned water," he told me, adding that the salient key sentences from the National Research Council report were particularly damaging. "Look, I don't know what the science is," he said.

"I just know that there was a scientist who said, 'This isn't a settled question yet.'" After that, Henderson just did what he had done for years, helping to make sure the public was aware of those insecurities. "The issues were polarizing," he told me. "I just knew how to express them in a way that maximizes the polarization."

Others put an even finer point on it. "The lesson learned from San Diego was you can't have the sewer authority as the spokesperson," said Patricia Sinicropi, executive director of the national WateReuse Association. She told me that the lead spokesperson on any potable water recycling project "has to be the water department—that was one of the big [San Diego] takeaways." In other words, no matter how articulate Schlesinger and Gagliardo may have been, they were still representing the sewage team, which can make for bad optics when talking about drinking water. "Cities should assume that projects altering precious water resources will alarm the public," reported one San Diego postmortem. "[The media] quoted key opposition leaders, sensationalized the fears expressed, and focused on the 'yuck factor' instead of the stability and benefits of the program."[43]

⌣

But on the most basic level, from a communications standpoint, the term *water repurification* lost out in a head-to-head battle with *toilet to tap*. The more officials ran away from the toilet-to-tap label, the more the media, and critics, pushed the term. Appellations like "water repurification" and "reservoir augmentation" were seen as dodgy public relations spin. In the wake of the San Diego meltdown, the water recycling industry has struggled to come up with a positive catchphrase that could rival the blunt simplicity of toilet to tap. "Water reuse" is the industry's favorite these days. There's also "water recycling," "water reclamation," and even "showers to flowers." One industry poll suggested that the term *purified water* may connect particularly well with audiences.[44] But nothing resonates quite like the punchy and alliterative toilet to tap.

Then there was the cheeky contest that the *Voice of San Diego* ran in 2007. The online news organization—which generally has offered fair and balanced coverage of water reuse—held a contest asking readers to nominate alternate names for water recycling. Their staff called it the "Come Up with a Funny Name for Sewage Recycling Contest."[45] The newsroom announced, "We here at *Voice* headquarters have been joking about the many monikers for water reuse." So the staff decided to award a travel mug to the reader who came up with the best new term "for indirect potable water reuse—the bureaucratic term for recycling sewage."[46]

Contestants were required to nominate "something new and funny," with extra points for alliteration, but no name calling and no four-letter words. To help inspire reader participation, the staff offered a quick starter list of terms already compiled: "water reuse," "sewage recycling," "reservoir augmentation," "toilet to tap," "feces to faucet," "sewage to spigot," and "backside to frontside." During the next twenty-four hours, a slew of sophomoric suggestions poured in, including "poop to soup," "toilet bowl to pie hole," "fecal sequel," and "quench with stench." And the winner was … "asses to glasses"—a label that makes "toilet to tap" sound good—and it all came from a publication known for serious and balanced coverage of water recycling.[47]

Contests like that, while humorous, don't help the water recycling movement—and again, they show the difficulties of maintaining control of the narrative. For many online news organizations like the *Voice*, there is an expectation to be hip, trendy, and yes, funny. But nothing kills a good joke like a bad drought, and drought would soon return to San Diego like never before.

So would the city's push for purified sewage.

CHAPTER 3

Orange County Sets the Bar

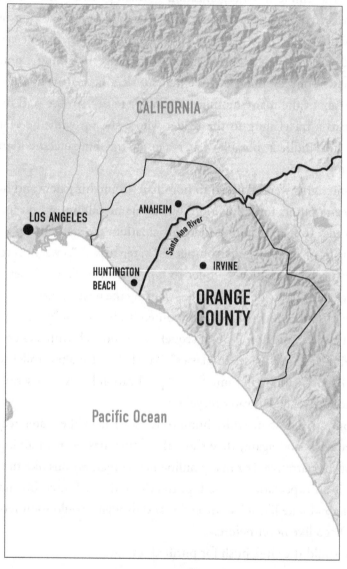

Orange County, California

San Diego's leaders had spoken. By rejecting water recycling so dramatically, they fed an anti-water reuse hysteria that threatened to spread far beyond the city. No place was more concerned about that than Orange County, just ninety miles up the road. Already a national leader in water reuse, Orange County had been investing—heavily—in water recycling for decades, and officials there watched San Diego's controversy with incredulous dread. They could not afford to be sucked into the city's toilet-to-tap melodrama. "It was such a disaster," remembered one official. "We were really concerned about the public relations."[1]

Orange County's storied water recycling program began in 1975 with something called Water Factory 21. It was the most futuristic water recycling initiative the United States had ever seen: water treatment with a twenty-first-century flair, hence the number "21" in the name. Orange County's groundwater had been under stress for years due to over-pumping and saltwater intrusion. But the county's groundwater was not always this troubled. At one time the aquifer was so robust that water was literally gurgling out of the ground. In 1888, the county was home to numerous natural springs and a three-hundred-square-mile "artesian zone" containing hundreds of free-flowing wells. As the county's population grew and agricultural irrigation swelled, the artesian zone shrank, contracting to just fifty-two square miles by 1923, when the water table was declining by two and a half feet per year.[2] Eventually the artesian zone would disappear altogether. Along the Pacific Coast, saltwater intrusion was even more disconcerting. Back in the 1880s, the burgeoning aquifer created so much pressure that it would push freshwater into the Pacific up through the ocean floor. The flow reversed when the groundwater dropped, however. Saltwater then seeped through the ocean floor into the aquifer, contaminating numerous wells and threatening a wider swath of Orange County's drinking water supply.

During the 1950s, officials came up with a plan of attack. They created spreading basins—shallow ponds with porous bottoms—that they

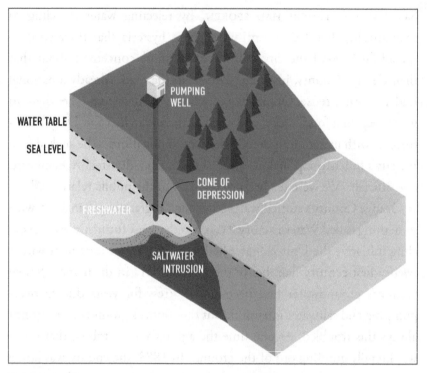

WATER TABLE

SEA LEVEL

PUMPING
WELL

CONE OF
DEPRESSION

FRESHWATER

SALTWATER
INTRUSION

Figure 3-1. Saltwater intrusion can occur in coastal areas when too much water is pumped out of an aquifer.

filled with imported water from the Colorado River. That water trickled down through the bottom of the ponds into the ground, gradually recharging the aquifer.[3] In the 1970s, injection wells were added so that water could be pumped directly into an underground barrier that had been engineered to keep saltwater intrusion at bay.[4]

Between the spreading basins and injection wells, Orange County's groundwater levels had improved markedly by the 1970s. Now, after decades of using Colorado River water to restore the aquifer, the plan was to use the Water Factory's purified sewage in the injection wells instead. The idea was to mix the recycled water with other water and then pump the blended concoction underground. The Water Factory's key technology was reverse osmosis, an expensive water purification technique to turn seawater into drinking water. But Orange County planned to use reverse osmosis to purify sewage instead. It was a bold strategy. Would it work?

Water Factory 21 captivated experts from around the world, many of whom traveled to Orange County to see water recycling in action. Before the effluent even hit the reverse osmosis membranes, it was subjected to several pretreatment layers, including lime clarification, recarbonation, and rapid sand filtration. All those processes had been around for years, but it was reverse osmosis that garnered expert attention. Locally, Water Factory 21 didn't create all that much buzz. Because the primary role of the injected water was to stanch saltwater intrusion, local residents didn't think much about the purified sewage being slowly blended with their drinking water underground. As a result, Orange County's citizens backed into the water recycling movement without completely grasping exactly what was happening below their feet.

Devoid of controversy, Water Factory 21 flourished. By the early 1990s, after years of data collection showed the process to be safe, health officials became so confident in the technology that Orange County was permitted to pump the purified effluent directly into the ground, without blending it with other water first. Just a few years later, in the mid-1990s, amid growing concerns about drought and the reliability of imported water from Northern California and the Colorado River, Orange County officials decided to dramatically expand their highly successful water recycling program. Building on decades of innovative research, success, and public acceptance, they quietly drew up plans for a water recycling facility that would nearly quintuple their volume from fifteen million gallons per day to seventy million gallons. It would be the largest program of its kind in the world and would transform Orange

County's water recycling program from a novel groundwater recharge effort to a major segment of the local drinking water supply.

Or so they hoped. Just as Orange County was about to ramp up its vaunted program, a similar initiative was disintegrating ninety miles down the road. Would San Diego's toilet-to-tap chaos spread north and drag down Orange County's world-renowned program with it? That was no small question. The problem was that Orange County, unlike San Diego, wasn't just dabbling in water recycling for the first time. It had grown very dependent on water reuse and was about to spend $352 million on a major water reuse upgrade. What would happen if San Diego's controversy prompted Orange County's citizens to suddenly rebel against water recycling? They had already been drinking it for years.

Rather than wait and see, the Orange County Water District launched a sweeping public relations offensive and hired Ron Wildermuth to lead it. Wildermuth was not your average PR guy. He had been the lead communication officer for General H. Norman Schwarzkopf, the highly decorated and charismatic US Army general who oversaw Operation Desert Storm in the early 1990s. Schwarzkopf proved to be a captivating and skilled communicator, mesmerizing the world with live press conferences and deftly narrating Persian Gulf combat videos for a global audience. As Schwarzkopf's front man, Wildermuth supervised fifty people who were feeding real-time war zone information to more than a thousand journalists around the world.

Working for Orange County was a very different gig. "The first thing I did, of course, was drink the water," he told me, adding that if he had any doubts about Orange County's purified sewage being "the best quality water the citizens could get, I wouldn't have done the project." He quickly proposed an ambitious community-outreach effort that would include hundreds of face-to-face briefings with key public figures, as well as an extensive slate of presentations to the general public. The plan was to target politicians, business leaders, and religious figures with one-on-one briefings, along with talks at social clubs, at chambers of commerce, to academics, and to doctors. He also proposed a host of large presentations reaching hundreds of average citizens. By going straight to the public with his messaging, Wildermuth hoped to control the narrative and overwhelm any bad publicity from San Diego or anywhere else.

Water District officials signed off on his plan, and that's when Wildermuth panicked. "I was afraid to speak in front of the public," he told me. "It was something I *really* had never done before." Sure, during his military career he had set up numerous large press conferences, but it was Schwarzkopf who took the podium. Wildermuth ran some large meetings, but those were with his fellow troops. Getting up in front of hundreds of people at a public hearing was very different. "Desert Storm was strictly *media* relations," he said. "With the water business, the idea was *community* relations.... I couldn't do it."

But he did. Somehow, he shook off the glossophobia and got to work. "We gave talks seven days a week—nights, mornings, evenings—everything," he remembered. "I mean we went to every leader in Orange County." He brought show-and-tell too—microfilters, reverse osmosis tubes, anything that would help make the case. Letters of support were key. Every time he made a new convert in one of his face-to-face meetings, he asked for a letter endorsing the program, and the officials obliged. With each letter, he would add that organization's logo to his PowerPoint deck, showing a widening base of support with each presentation. That, in turn, led to more converts. He ended up with hundreds of letters and an endless stream of logos in his presentation.

The numerous sessions with average citizens were also integral. Wildermuth ran newspaper ads inviting people to "workshops," where he would give the same presentation delivered to community leaders. "Concerned about future water in Orange County?" the advertisements asked. "Learn more! ... Voice your opinion!"[5] Because Orange County was becoming increasingly diverse, the district hired native speakers to conduct sessions in Spanish and Vietnamese as well. The talking points have remained the same in Orange County for years. (1) Imported water from the Colorado River and Northern California was unreliable and had already been cut off or reduced during droughts. (2) Orange County needed alternative water sources, and purified sewage was a local, drought-resistant supply that officials could predict and control. (3) Water reuse was good for the environment; it meant that less sewage would be discharged to the ocean. (4) Reuse was expensive, but it would become cost competitive in the future as the price of imported water rose. (5) Most important of all, it was safe.

During these sessions Wildermuth pounced whenever things veered off message. "If anybody said 'toilet to tap,' we said, 'Whoa, wait a minute! That's absolutely wrong!'" he told me. "We just hit it—pow!—head-on." In a remarkable extra step, he solicited written endorsements from the general public as well. He would hand out postcards and ask everyone in the room to sign them, endorsing the program. Cards poured in by the hundreds.

During our interview, Wildermuth rattled off his spiel like he was still on the job, making it clear that the new treatment plant was much more technologically sophisticated than Water Factory 21. The first step in the new facility was microfiltration, which would force the wastewater effluent through the walls of straws containing pores that were "three hundred times smaller than a human hair," he said, screening out suspended solids, bacteria, and protozoa in the process. The next step was reverse osmosis, which forced the water through membranes, removing viruses, pharmaceuticals, and dissolved minerals. That was followed by a third wave during which the already purified water would be blasted with high-energy ultraviolet light and splashed with a dose of hydrogen peroxide to disinfect it even further and destroy any remaining trace organic chemical compounds. The result would be something akin to distilled water, safe enough to drink at home—but that's not where it went. Instead, the water was either injected into the saltwater barriers or sent to spreading basins. From there, it would take months or even years for the purified sewage to migrate underground to drinking water wells, not to mention reach household taps. In the meantime the recycled water would mix over and over again with other water that had seeped down into in the aquifer naturally. The new facility was not only bigger than Water Factory 21, it was better. "Inch by inch," Wildermuth said, "we got huge support."

One surprise emerged from the community meetings. To Wildermuth's amazement, people did not realize that Water Factory 21 had been blending purified effluent with their drinking water for decades. "Nobody knew about it," Wildermuth said. That gave him a significant PR advantage. How could they oppose an expanded water recycling facility—that would employ more sophisticated purification layers than Water Factory 21—when everyone in the room had been drinking the Water Factory's product for decades? "We did use the experience of Water Factory 21—saying we'd done it, and it's safe, and it's been going on for years—but we had to bring it to their attention."

Another tactic proved more challenging. Wildermuth wanted to bottle the Water Factory's purified effluent and hand it out at his presentations, but the California Department of Public Health forbid it. "So, I did it illegally," he confessed. He brought the recycled water to his briefings in a thermos. Then, toward the end of his talks, he poured a cup and quaffed it down. He would also pour small cups for a handful of people to sample for themselves. "By the time we got done talking about it," he said, "it was so far from sewer water that it broke the connection."

It showed. At one of the first official public hearings about the new water recycling facility that was held at the Orange County Water District headquarters, audience support was overwhelming. "Our analysis shows this is probably one of the most innovative projects to come forth in Orange County in a long time," said Tony Aguilar with the Orange County Hispanic Chamber of Commerce. Then an environmental advocate from the Surfrider Foundation went even further, saying that recycling just seventy million gallons of wastewater effluent per day was not enough and that the county should recycle even more. Public support at the meeting was so formidable that the official chairing the session was beside himself. "We're trying to get negative input, but we can't."[6] During Wildermuth's tenure, Orange County Water District officials gave more than twelve hundred presentations about their new program—most of them by Wildermuth—including an extraordinary flurry of talks during his first few years on the job.

PR challenges eventually cropped up, but even in those instances, Wildermuth's team kept them from spiraling out of control. Bad news that might have been the beginning of the end in places like San Diego failed to catch hold in Orange County. In May 2000, Orange County abruptly announced that it had closed two wells after discovering contamination from NDMA, an organic chemical and probable carcinogen.[7] What made the incident even more damaging is that the contamination was an inadvertent result of Orange County's sewage recycling process. (NDMA can be an unintended by-product of treating wastewater *and* drinking water.) Officials shut down the plant until the problem was resolved.

They also reassured the public that the detections were in minute amounts and the water supply was safe. They set up an NDMA hotline and encouraged people to call. But they also emphasized that NDMA was a ubiquitous probable carcinogen found in many processed foods at much higher levels than had been detected in the groundwater. "You probably got more of this chemical over the [Memorial Day] weekend," Wildermuth was quoted as saying, "[by] eating hot dogs and drinking beer."[8] The wells reopened after additional onsite treatment systems were added to scrub out the NDMA before the water was sent to public pipes. No backlash occurred.

A similar scare happened two years later, in 2002. This time nine Orange County wells were closed due to contamination from another chemical, 1,4-dioxane, and again it was determined that the contaminant entered the groundwater partly as a result of the water recycling process.[9] "The low point was really the discovery of NDMA and then the discovery of 1,4-dioxane [and] that we had, in fact, added to the contamination of the groundwater basin," remembered Mike Wehner, who was in charge of water quality at the Orange County Water District at the time. "Our principal job was to protect and maintain the groundwater basin," he told me. "To find out that you've been contributing to contamination in the groundwater basin is horrible."

Water District officials responded to the 1,4-dioxane incident in the same low-key manner, reassuring the public that the water was safe, and announced that they would use the same solution: installing treatment systems at the wellhead to remove the chemical before it entered the drinking water system. To prevent further contamination of the groundwater, they also added ultraviolet and peroxide treatments to the tail end of Water Factory 21 to remove the chemicals as the water left the plant. On the PR front, officials emphasized that if citizens were really worried about 1,4-dioxane, the minuscule amounts detected in the wells were the least of their concerns. Local media coverage reported that the chemical is found in "shampoo, dishwashing soap, baby lotion, hair lotions, bath foam and cosmetic products ... [and] exposure to this compound can be by inhalation, orally, or through the skin."[10] After a brief flurry, the 1,4-dioxane incident blew over too.

In the end, the most serious opposition to the new water recycling facility was about money, not water quality. A few outspoken community leaders said that the new plant was just too costly and that the water crisis was not yet urgent enough to justify the expense. Orange County water officials disagreed, arguing that infrastructure of this sort took years to build and that now was the time to act. They admitted that the price was steep but insisted that the investment in water security was worth it—especially for a predictable, local, drought-resistant supply. By the time the new plant was built, the water situation would be plenty urgent. What's more, they reminded people that as the cost of imported water continued to rise, water recycling would become increasingly cost-competitive over time. Despite the cost concerns, most people favored the project, and the Orange County Water District broke ground on the new seventy-million-gallon-per day (mgd) facility in 2004.

⌣

While Wildermuth worked to win over the public, Water District engineers were preparing to build the most ambitious water reuse facility the world had ever seen. Mike Markus emerged as a key figure during this time. A civil engineer and low-key California native, Markus spent the first decade out of graduate school building water and wastewater plants around the state. He joined the Orange County Water District in 1988 and stayed on for more than thirty pioneering years. In the early 2000s he was put in charge of building the new water recycling operation. From the beginning, that effort was seen by the Water District's board as just the first phase of a multidecade vision of water reuse expansion. After building the 70 mgd plant, the long-term plan was to increase capacity to 100 mgd a few years later and then boost production yet again to 130 mgd a few years after that. These were unprecedented volumes in potable water recycling, and they would keep Orange County at the vanguard of the global water reuse movement for years to come. "That was certainly a defining moment in the history of the Water District," Markus told me. "We were introducing this new technology, and not at five or ten mgd, but at a seventy-million-gallons-per-day scale. Our board went all-in very early on."

Vision is one thing, but Markus had to build it. While his immediate task was to erect a 70 mgd plant within a snug footprint, he also had to carve out space for the two additional 30 mgd upgrades that would be added later. It took four years to complete the first phase, and space was so tight that Water Factory 21 had to come down. Since no one had ever built anything like it, Markus and his team were under no small amount of pressure. "We threw the switch in January of 2008," he told me, with characteristic understatement, "and you know, everything worked." They called it the Groundwater Replenishment System, or GWRS—not as sexy as Water Factory 21, but it would prove to be even more transformative. Once again, experts flocked from all over the world to see it. As with Water Factory 21, the plan was to inject GWRS water into the saltwater intrusion barriers, but also to send it to the spreading basins for aquifer recharge.

Then, in February 2013, Orange County was hit with another contamination curveball: an unknown industrial customer illegally dumped an enormous amount of acetone into the county's sewer system.[11] Acetone is a colorless solvent found in nail polish remover and paint thinner, and it has numerous industrial applications as well. The ditched acetone passed through Orange County's sewage treatment plant undetected before continuing on to the GWRS for purification. For precisely this reason, the GWRS plant is riddled with real-time sensors that monitor the water as it moves through the facility, constantly searching for contaminants. The alarms went off. The sensors couldn't tell exactly what the contaminant was, only that it was bad. "We could see we had a spike coming in, and then we could see the spike continuing," recalled Wehner, head of water quality. "We knew it was volatile. We had to deal with it."

Wehner told his staff to grab a sample for analysis. But the testing took time, and during the next twenty-four hours the real-time monitoring showed the spike continuing to climb. When the sample came back as acetone, Wehner said, the spike was declining. They were relieved to learn it was acetone, Wehner told me, because they knew acetone would dissipate in the groundwater system without impacting the water supply. "We came close to shutting down but didn't," Wehner said. "The spike had gone away." The incident was alarming, but the acetone level was never high enough to violate the plant's water quality permit, in part, Wehner said, because there is so much water volume moving through the facility that the spike, while detectable, was already quite diluted.

But how did the acetone get through all those layers of treatment? Reverse osmosis removes *almost* everything from sewage, but not acetone, which is why Orange County's water recycling plant has the ability to offload the incoming water or shut down completely when certain contaminants are detected. "It was kind of an 'Oh My God' moment," admitted Markus. "We always say that [reverse osmosis] takes care of most of the contaminants. However, there are some low-molecular-weight organics that do get through."

This incident is exactly what people worry about with sewage purification. They fear that something dangerous will contaminate their drinking water, harming them or their families. Such incidents show the importance of having real-time sensors in water recycling plants that alert team members about problematic substances in the incoming stream of treated effluent. Similarly, water reuse plants are equipped with diversion systems that can immediately redirect contaminated water offline, sending it out of the plant. There is also mounting pressure in the industry to install even more monitoring equipment, not just in the water recycling plant, but also in the sewage treatment plant that feeds the effluent to these reuse facilities. Increasingly, places like Orange County will be expected to have a better handle on what's happening, in real time, throughout their "sewershed"—the entire geographic watershed from which a community's effluent is collected. Sewershed monitoring became common during the COVID-19 pandemic to track the illness's spread.[12] (The coronavirus is detectable in human waste—even from people who are asymptomatic—but the virus is also very easy to neutralize in water recycling plants.) Sewershed monitoring can also be used as an early alert system for illegal dumping too, like the acetone incident in Orange County. "You're getting the whole sewershed kind of aligned with the reuse program," explained G. Tracy Mehan III of the American Water Works Association. That said, sewage purification is still a mind-over-matter enterprise, dependent on the public's trust, and if similar incidents were to occur repeatedly at a reuse facility, public confidence could erode.

To Markus the take-home message is that people should have more confidence in the system, not less. The spike was detected, it was short-lived, and it was so diluted that the plant's permit was never violated. The water supply was never threatened. "The story is that the system worked," he said. "The monitoring found it, [and] that's really the only anomalous experience we've had." Once the spike passed through the plant, it went on into Orange County's groundwater where it biodegraded in the aquifer and became undetectable.[13] It never showed up in the district's extensive groundwater monitoring. California officials say that the acetone never got close to entering the drinking water system.[14] "That's one of the benefits of using groundwater augmentation," Markus said. "If any nonspec water got out, you've got the travel time in the groundwater basin to be able to handle that."

As Markus pointed out, the acetone incident highlights the benefits of discharging purified wastewater into an aquifer first, before it is withdrawn and treated again for drinking water. That way the groundwater serves as an "environmental buffer" between the recycled water and the public. In the rare event that contaminants, like acetone, sneak through, they are buffered in the groundwater system. This kind of water recycling is known as indirect potable reuse because the purified effluent is delivered to drinking water customers indirectly by passing it through groundwater. That is very different from direct potable reuse, which is more rare, more demanding, and more heavily regulated because it discharges purified sewage directly into the drinking water system, without passing through an environmental buffer.

Because no permit was violated during the acetone incident, Orange County officials did not notify the media, but they did file a report with the state and openly talked about the incident at water conferences and professional gatherings. This book is the first time that the acetone incident has been publicly reported outside conferences or academic papers. "Now I'm at conferences and I'm hearing about [Orange County's] 'acetone incident,'" Markus said with a smirk, adding that "we could have just as easily ignored it, [but] it was totally the right way to handle it." Markus told me that his team tightened their protocols after the acetone incident. The new procedures include a requirement to shut down the plant if a similar problem were to occur.

In the years that followed, the water recycling program continued to expand in stages, following the district's original plan. In 2015, under Markus's leadership, the first 30-mgd upgrade of the GWRS went online, which bumped up Orange County's water recycling capacity to 100 million gallons per day. Over time, the Water District's campus slowly evolved into a thriving water-recycling tech hub. Reverse osmosis manufacturers and other water treatment vendors regularly approached Markus for permission to road test gadgets in the new plant, something that began back in the days of Water Factory 21. The district formed a research and development arm and staffed its lab with twenty-five in-house chemists who spent a lot of time testing the groundwater as well. The R&D–vendor collaborations gave Orange County a front-row seat for new innovations, helping ensure that its own equipment was state of the art. "We have the capability to full-scale test a lot of different technologies," Markus said. "If we can use this as a test-bed to help prove out technologies, that is our contribution to help move the science along."

In early 2023 the final 30-million-gallon tranche of the Ground Water Replenishment System came online, topping out Orange County's production at 130 million gallons per day.[15] That milestone not only maintained Orange County's status as the largest producer of potable recycled water in the world, but it also became the first utility to recycle 100 percent of its available effluent. "We're tapped out," Markus told me. "That's all we can do."

Today Orange County reigns as the leading potable water recycling success story in the world. International visitors still come knocking, tour groups still roll through, and awards continue to pour in. The Water District's success has had an enormous nationwide influence on everything from sewage purification technology to regulations, communication strategies, and public acceptance, making Markus one of the most highly regarded figures in the industry. During more than thirty years at the Water District, he oversaw the construction of all three phases of the GWRS, finishing each without drama. "The system works," he told me. "You don't have to worry about the technology. This multibarrier process *works*." Other communities have built potable water recycling facilities in the United States, but none of them come close to Orange County's in size. Still others have pledged to construct potable reuse programs that are larger than Orange County's, but none of those are close to being built. "Orange County is the world leader in indirect potable reuse and groundwater recharge," said Felicia Marcus, former chair of the California State Water Resources Control Board. "I think they've provided a model that San Diego and LA and others are going to follow." And they are.

CHAPTER 4

San Diego Bounces Back

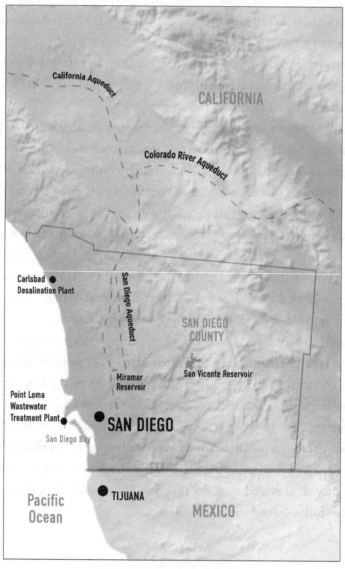

San Diego

IN THE EARLY 2000S, as Orange County was breaking ground on the most innovative water recycling plant in the United States, San Diego remained mired in water conflict. In the wake of the city's historic vote to kill its potable water recycling program, two new protagonists had moved to town. These young, talented, environmental attorneys were hungry to protect the ocean and determined to revive the city's water recycling effort. They had the benefit of arriving just as San Diego's toilet-to-tap war was ending, avoiding the scar tissue that clung to those steeped in the recent fight. But they strolled onto a water recycling battlefield that was riddled with professional burnout, broken dreams, and shattered egos. Morale was so low after San Diego's 1999 city council vote, remembers Marsi Steirer, a longtime Water Department employee, that people were afraid to even say the words *repurified water.* "It was almost like a cussword or something," she told me. "You didn't talk about it.... It was 'the project that shall not be named.'"

The environmental lawyers carried none of that baggage. Marco Gonzalez was a San Diego County native passionate about surfing and fishing. He had gone off to the University of California Santa Cruz for undergraduate work and earned his law degree at Lewis and Clark College in Portland, Oregon, before returning home. A brilliant lawyer and searing debater, he joined the local office of the Surfrider Foundation and quickly moved into the top position. As he settled into San Diego's environmental community, Gonzalez eventually crossed paths with Bruce Reznik. Reznik was a Los Angeles native and Berkeley grad with a law degree from the University of San Diego who ran the local Baykeeper office. A skilled communicator and strategist, Reznik took over San Diego's Baykeeper office where Gonzalez ended up working as in-house counsel—in addition to his Surfrider duties. They became fast friends, bonding over a mutual concern about San Diego's sewage management and how those practices impacted ocean health. "I'll tell you what, as a thirty-year-old lawyer taking on the establishment over billions of dollars, it was heavy," Gonzalez recalled. "But Bruce and I felt like we could accomplish just about anything."

Their legal strategy focused on the Clean Water Act, or rather the Faustian bargain that San Diego had cut with the US Environmental Protection Agency (EPA) over that 1972 law. A key goal of the act was to stop cities nationwide from dumping untreated or undertreated sewage into rivers, lakes, and oceans. Under the act, the EPA forced communities to upgrade their sewage treatment systems, often with generous federal subsidies. Because San Diego's sewage was discharged into deep ocean waters, miles offshore, the city repeatedly and successfully pushed for a Clean Water Act waiver, arguing that the sewage rarely if ever escaped above a stubborn thermocline hundreds of feet below the surface.[1] These waivers are extremely rare, and San Diego was one of the last cities in the United States to be allowed to dump partially treated sewage into a surface waterbody.

The EPA waiver had to be renewed every five years, however, which gave Gonzalez and Reznik an opportunity to challenge the renewal in court, and with the EPA. That's what they did in 2002. As the litigation flew, snarky dueling continued in the newspaper. One op-ed blamed "libertarians and politicians" for killing San Diego's water recycling program and said that attempts to brand water recycling as "toilet to tap" was like calling milk "teat to table."[2]

Bickering aside, the litigation had impact. It led to more than a year of complex and fraught negotiations between Gonzalez, Reznik, Audubon, and the Sierra Club on one side, and city officials on the other. In 2004 the city agreed to a settlement that included three concrete deliverables: conduct more ocean monitoring, invest in additional sewage treatment technologies, and—most significantly—produce an in-depth feasibility study on water recycling.[3] In exchange, the environmentalists stood down on the EPA waiver, and San Diego's moribund water recycling program began crawling back to life. The public remained skeptical, however. Polling showed that 63 percent of county residents were overwhelmingly opposed to potable water recycling. It was fine to spread it on lawns, but few wanted to drink it.[4]

Despite that opposition, other changes were helping revive San Diego's water reuse movement. Environmental advocates and water recycling proponents were elected to the city council, especially Scott Peters and Donna Frye, who turned out to be passionate advocates for water reuse. A citizen panel voted unanimously to support potable water recycling as well.[5] The media coverage also changed. Kathryn Balint, whose hard-hitting *San Diego Union-Tribune* articles had defined the city's water recycling debate for years, was replaced by Mike Lee, a water reporter from the *Sacramento Bee*. Balint's tough stories usually included the wording "toilet to tap" and tended to regularly quote the same water reuse critics, especially former city councilman Bruce Henderson. In contrast, Lee's coverage was more balanced, mentioning toilet to tap less frequently, and called on a broader spectrum of voices. The Water Department changed too. Taking a cue from Orange County, San Diego officials went on the offensive. Steirer started sending scolding letters to the editor that reinforced the city's talking points and castigated report-ers for using the term *toilet to tap*. "It is time for the media to do away with this misnomer," she wrote, "it is misleading, inaccurate and inhib-its thoughtful public dialogue on water recycling."[6]

That's when San Diego officials did what they should have done a long time ago: they traveled to Orange County for advice—especially on public relations. "Orange County water officials watched in hor-ror six years ago as politics and public revulsion drowned San Diego's first attempt at turning sewage into drinking water," the *Union-Tribune* reported. "With jaws clenched, they launched an all-out public rela-tions campaign to preserve their own infant project." Now San Diego officials were "likely to adopt a communications strategy similar to Orange County's."[7]

In 2006 the city released the water reuse study that the environmentalists had fought for. The lengthy report, led by Steirer, vetted a slew of water recycling options, including potable and nonpotable projects. The report declined to endorse a specific proposal, but it declared all of them "feasible," even those producing potable recycled water. The report warned, however, that the viability of these projects would depend on the city's commitment and ability to garner support "through an extensive public involvement program"—like Orange County's.[8] Steirer remembers traveling to Orange County several times, where she found the staff to be welcoming and gracious. "We really relied closely on Orange County," she said. "Our public outreach plan was in large part based upon the parameters of what Orange County did." That plan included obtaining letters of support; holding briefings for officials, the media, and the general public; and conducting stakeholder interviews, focus groups, and public opinion polls. "We definitely were following their playbook," she said.

⌐

Drought returned in 2007, threatening cutbacks to San Diego's imported water supplies—possibly for years. Historically, California has been hounded by drought, but starting in the early 2000s each dry spell was seemingly more severe than the last. As each drought settled in, insecurity built over the reliability of water imports from Northern California and the Colorado River, and support for water reuse in San Diego began to rise. What's more, the city council emerged as a strong champion. In 2007, in a major step in the wake of the water reuse study, the council approved a $10 million potable reuse pilot facility to produce one million gallons of purified sewage per day.[9] The public could tour the small plant and become more comfortable with water recycling technology. But the *Union-Tribune's* editorial board remained adamantly opposed to potable reuse. The paper repeatedly published scathing editorials that referred to water recycling as a "boondoggle" or a "mindless" multimillion dollar "toilet-to-tap scheme" that should be "scratched for health and safety reasons."[10]

Then, in 2011, something astonishing occurred. The screaming opposition to potable water recycling that had reverberated from the *Union-Tribune*'s editorial page for years suddenly reversed course. In one of the most remarkable examples of an about-face ever to occur at a major metropolitan daily, the newspaper published a historic crow-eating editorial that retracted a decade of scathing criticism and fearmongering. Under the headline "The Yuck Factor: Get Over It," the paper heralded San Diego's new million-gallon-per-day water recycling demonstration facility, explaining that if it proved safe and affordable, the city would build a much bigger plant. "Frankly, there's not much to demonstrate," the paper admitted, pointing out that water managers in Orange County, Singapore, and other places around the world had already proven the technology to be safe. "Similar efforts in years past were dubbed by critics, including this editorial page, as 'toilet to tap,'" the paper said. But in a courageous mea culpa, the *Union-Tribune*'s editorial board said that it had "come to accept" that recycled water "would likely be the purest and safest water in the system." Yes, many people will still harbor a "yuck factor," the paper admitted, but "in our view, it's time to get over it."[11]

"That was huge," Gonzalez told me—and not just for environmentalists. Down at the city Water Department, people couldn't believe it. Steirer said that "everybody was jubilant and jumping up and down." But the editorial board's pivot reflected a palpable shift in perspective about water reuse that was sweeping through San Diego. Years of drought, water restrictions, climate anxiety, and the media's nonstop drumbeat about water insecurity had softened local opposition to potable reuse. People were coming to realize that there weren't many water options left. A broad coalition of water recycling advocates emerged to back up Gonzalez and Reznik, including the biocom industry, builders, and labor. Diversified supply was all the rage, especially if it helped wean San Diego off imported water. An expensive new seawater desalination plant was planned north of town, but it would only meet 10 percent of the region's water demand. Polls that once showed opposition of 63 percent to potable reuse now showed 67 percent in favor. "Backers of water recycling see a rising tide of support," the *Union-Tribune* declared. "Think of it as the dawn of an era in which every drop of sewage is scrutinized for possible reuse."[12]

Things changed nationally too. The National Research Council, which had released the problematic report in 1998 that called projects like San Diego's an "option of last resort,"[13] issued a new study in 2012 reversing prior conclusions. It now declared purified effluent to be as clean or possibly even cleaner than traditional water supplies. "Wastewater reuse is poised to become a legitimate part of the nation's water supply portfolio," said R. Rhodes Trussell, chair of the committee that wrote the new report.[14]

What changed during that fourteen-year timespan? David Sedlak is the author of the book *Water 4.0*, which includes a chapter on reuse, and he served on the committee that produced the 2012 study. He said that water recycling technology and practice—in Southern California and even Singapore—dramatically improved between 1998 and 2012 to the point where the council believed that it was time to fully endorse potable water recycling. "Between 1998 and 2012," Sedlak told me, there were "a lot more years of experience and data in places like Orange County where these treatment plants were operated under great amounts of scrutiny.... I think that built a lot more confidence among the research community as well as among practitioners."

Several people interviewed for this book attributed the hesitant 1998 National Research Council language to one scientist, Daniel Okun, a professor at the University of North Carolina. Okun, who served on the panel that produced the 1998 report and passed away in 2007, was well known in the water-research community as a potable reuse skeptic.[15] The extensive new health and safety data from Orange County and elsewhere made Okun's minority opinion seem all the more marginalized, helping pave the way for the council's do-over in 2012.

Throughout this period San Diego was still coasting along on that rare Clean Water Act waiver. When it came up for renewal once again, Gonzalez and Reznik threatened to drag the city back into court unless officials doubled down on potable reuse. That prompted the city to agree to yet another major water recycling study. If the earlier study was designed to convince citizens that potable reuse was possible, this second study showed how a full-fledged water recycling program could be built, producing up to one hundred million gallons of recycled water daily. For the second time in a decade, litigation—or the threat of it—from Gonzalez and Reznik played an enormously important role in guiding the future of water recycling in California's second largest city. "We now have a blueprint that ... will hopefully change the way we view sewage," Gonzalez said after the second report was released.[16]

During this time, scientists grew increasingly alarmed about drought in the Southwest, eventually calling it the worst the region had seen in twelve hundred years.[17] Experts would end up coining a new term, *megadrought*, to describe how climate change was transforming the regional environment.[18] San Diego's support for potable water reuse continued to rise, hitting 73 percent,[19] the highest in city history. "Water shortages," quipped the *Voice of San Diego*, "are scarier than irrational fears about pee."[20]

Then, in a move that seemed unfathomable fifteen years earlier, the city council took the historic step in 2014 of approving a new signature potable water recycling program.[21] They called it "Pure Water San Diego." The ambitious $3.5 billion project would involve three different water recycling plants spread throughout the city. It would take decades

to build them, but they were slated to produce eighty-three million gallons of purified sewage by 2035—one-third of the city's overall supply.
It was the largest capital investment in the city's history, and all of it
would go toward water reuse.[22] "Pure Water is a cost-effective innovative
project," Mayor Kevin Faulconer proclaimed. Yes, he admitted, it was
expensive, but "the cost of doing nothing is even higher."[23]

For long-time observers, there was an irony in the Pure Water plan.
It used the same setup that was rejected in 1999: purify effluent in a
complex multitiered process, send it to a reservoir where it would mix
with other water for a while, and then treat it again before piping it
to homes. This time San Diego planned to send the recycled water to
an even smaller reservoir than in 1999, which meant that the recycled
water would spend even less time mixing before being piped to homes.
None of that fazed the public this time around. "It's really important to
understand that water is something that is precious, and should be used
and reused over again," Steirer told me.

Momentum lagged a bit as legal and construction issues delayed the
project for seven years.[24] Finally, in 2021 construction on the first phase
of Pure Water began, with projections now showing that recycled water
could eventually make up nearly 50 percent of the city's water supply
by 2035.[25] "It's just wonderful for the region, for the city, for water
security," said Steirer, who is now retired from the San Diego Water
Department. "It will provide the city with a level of water security that
it didn't know previously."

Pure Water represents a historic success for San Diego, but it was an
extraordinary accomplishment for the city's environmental community
as well—especially for Gonzalez and Reznik. Scores if not hundreds of
people are responsible for the city's water reuse about-face. But Pure
Water might never have come about if not for the novel legal strategy
employed by Reznik and Gonzalez. "Their involvement, I believe, was
key to bringing the concept of potable reuse back," Steirer told me.

Many others agree. "Marco and Bruce are masters of public relations and know how to get attention," said one water official in San Diego who spoke on the condition of anonymity. Using environmental litigation, or the threat of it, to leverage the city's Clean Water Act waiver was a "critical factor" in San Diego's water recycling revival, the official said, adding, "Their savvy in using the Clean Water Act to make this a priority is ultimately responsible for it happening." Reznik, who has since moved on to lead the Waterkeeper office in Los Angeles, called his work in San Diego "one of my proudest achievements" and said, "I think anybody in the know would say Pure Water, or what became Pure Water … only came onto people's radar because we sued over this waiver."

The philosophy behind the environmentalists' legal strategy proved crucial. Rather than use the Clean Water Act to force San Diego to upgrade its sewage discharges incrementally—to what is known as "secondary treatment"—Gonzalez and Reznik helped convince the city to dramatically reduce its sewage discharges by tens of millions of gallons per day. By converting the discharges into recycled drinking water, not only would the purified effluent provide a new, predictable, local, drought-resistant water supply, but it would reduce the amount of sewage discharged to the ocean as well.

Gonzalez still remembers when the idea hit him. He was sitting on his surfboard, undulating in the ocean, waiting for a wave. That's when he realized what a waste it would be to spend billions on a sewage plant upgrade—because millions of gallons of moderately cleaner sewage would still be discharged to the sea. Instead, he realized, those funds could be directed toward a sewage purification plant that would prevent millions of gallons of sewage from entering the ocean at all. "We were going to spend $2 billion to upgrade that plant to secondary treatment," he said, "and I was like, 'Why would we do that?'" At first, other environmental groups struggled with this approach. The knee-jerk reaction was that the EPA waiver had to go; San Diego's sewage plant should no longer be permitted to avoid the water quality upgrades that the vast majority of other major water treatment facilities in the United States had long-since completed. But in a touch-and-go moment with colleagues from other environmental groups, Reznik and Gonzalez managed to win them over.

Gonzalez has fond memories of his early conversations with Reznik about the idea. "Bruce and I would sit around, and I'd be like, 'Dude, I've got an idea, tell me how crazy you think this is: What if we said you don't have to build secondary [treatment] but instead we're going to hold your feet to the fire to do toilet to tap?' And we'd sit there and talk ourselves into this crazy idea, and we'd be like, 'Dude, I think we can do this!'" Today Gonzalez continues to surf and fish in the waters off San Diego. "That's still what I do in my free time, when I'm not with my family." In the end, it was all about helping San Diego residents embrace a paradigm shift in the way they think about wastewater. "What I really wanted to do was reframe the way we look at water in the arid Southwest," he said. "I wanted to reframe how we look at sewage."

CHAPTER 5

Future Water in Virginia

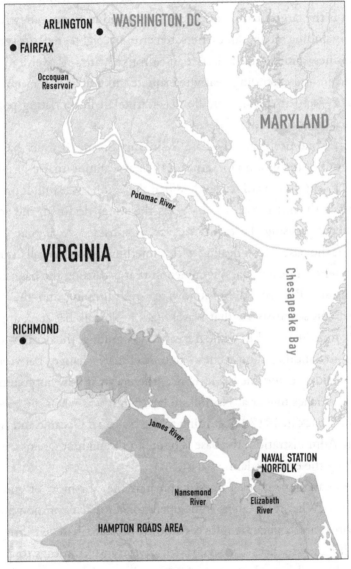

Virginia

NAVAL STATION NORFOLK IS A SIGHT TO BEHOLD. The view from atop the base's tallest building shows a striking array of docked warships stacked up for miles along Virginia's southern coast. Many piers are empty too, representing the numerous vessels out patrolling the high seas. It is the largest naval base in the world and is home to nearly sixty ships, including six aircraft carriers, ten cruisers, eighteen destroyers, ten submarines, and eighteen aircraft squadrons. Spanning twelve miles of waterfront and extending nearly ten square miles inland, the enormous complex—and its formidable flotilla—strike an intimidating pose for America's maritime adversaries.

There's one problem. The base is sinking. So much water has been pulled out of the ground underneath the naval station that it has slowly started to slump toward sea level. When the aquifer was full, it propped up the earth underneath the base. As the water declined, the aquifer compacted, causing the earth's surface to subside.[1] As if that weren't enough, the base is also dealing with something scientists call "isostatic adjustment," a postglacial phenomenon that is causing the base to sink even more.[2] Between the groundwater overpumping and the glacial adjustment, scientists say that Naval Station Norfolk has sunk approximately nine inches.[3] But while the earth was sinking, the ocean was also rising—by another nine inches—due to climate change. Between the land subsidence and the sea level rise, the ocean is now approximately eighteen inches higher at Norfolk today than it was when the base was first built back in 1917.[4] According to the National Oceanic and Atmospheric Administration, the base is home to the highest rate of sea level rise along the entire Atlantic Coast.[5]

The changing sea level doesn't threaten the warships, of course—they float. What it threatens is the operational capacity of thousands of uniformed and civilian employees who work at the base, ensuring that Norfolk's brawny fleet is stocked with everything from eggs to ammunition. Now, during the highest tides, employees drive to work through seawater. When hurricanes and other storm surges hit, large swaths of the base are inundated, making it even more difficult to tend the fleet. Thanks to climate change, Norfolk is the second most vulnerable port in the United States, after New Orleans.[6]

That's where Ted Henifin enters the picture. Affable and avuncular, Henifin is prone to wearing pink shirts and ties—although not at the same time—and vibrantly colorful socks too. He is the visionary behind a renowned sewage recycling project in southeast Virginia that may stop Norfolk from sinking while also helping clean up Chesapeake Bay—and potentially saving a billion dollars in the process. Here in the swampy lowlands of southeast Virginia, which get nearly four feet of rain per year, Henifin's brainchild proves that sewage recycling is not just a water supply solution for the desert Southwest—it can be an extremely valuable water source in wet areas too. Henifin's project also shows that water reuse can help the environment, save money, and—in this particular case—lend a gentle hand to national security as well.

⌣

Henifin grew up in a Navy family, spending most of his life in Virginia. After earning a civil engineering degree, he spent the next twenty-four years on public works projects for the military and local government. In 2006 he became general manager of the Hampton Roads Sanitation District, which is not your average sewer system. The wastewater utility treats the sewage of nearly two million people spread over a sprawling three-thousand-square-mile territory along the southern portion of Virginia's Atlantic Coast. Its unusual service area spans the mouths of three major rivers just at the point where they enter Chesapeake Bay. It is considered to be the most complex sanitary district in Virginia—by far. Most sewer systems operate primarily by gravity, but because Hampton Roads' service area is so flat, the majority of its sewage must be pushed through pressurized pipes, making it an even more complicated system.

The local economy is dominated by the Navy and other ocean-oriented interests. "Our whole tourism history is based on having this beautiful Chesapeake Bay," Henifin told me. "We have beaches on the bay, we have beaches on the ocean, a lot of recreational fishing, boating, sailing, oystering, crabbing … all that."

Henifin arrived at a rocky time in Hampton Roads' history. Shortly after he started, the US Environmental Protection Agency (EPA) hit the sanitation district with a punishing $3.2 billion water quality administrative order. The EPA was demanding upgrades to prevent sewage overflows that were ending up in Chesapeake Bay. Henifin saw the EPA's order as a misguided solution that would force him to fix problems that only existed in some parts of his service area while not adequately addressing pressing issues elsewhere. That prompted a multiyear series of negotiations with the federal government. Those negotiations led to a consent decree implementing a series of more nuanced sewer-system upgrades that would deliver targeted water quality benefits while also cutting the administrative order's price tag from $3.2 billion to $2 billion.

That was a huge success, but Henifin hungered for more. He realized that once these new, expensive sewage treatment improvements were made, in coming years the government would likely demand additional upgrades—followed by requests for more tweaks in the years after that. "We had actually done this fifty-year scenario-planning exercise," Henifin told me. "Every scenario we looked at basically had us making major upgrades to our plants about every five years." He decided that, in the long run, the continuous cascade of additional improvements would be expensive for ratepayers and disruptive to his team—bringing only incremental environmental benefits. Instead, he considered completely overhauling the philosophy behind how the sanitary district functioned.

The secret, as he saw it, was in water recycling. If he turned his sewage into drinking water, it would be better for the environment and he could finally get the regulators off his back. "Let's just leapfrog to drinking water," he told me. "They can't ask us to do more than that." It was the same conclusion that environmental groups had made in San Diego—that water recycling was a much better alternative than continuing to discharge treated sewage to the ocean. But in Virginia, the push was coming from the sanitation district itself, not the environmental community, and the move was driven as much by cost-cutting and convenience as it was about environmental health.

That's how Henifin's project, Sustainable Water Initiative for Tomorrow, or SWIFT, was born. SWIFT is a groundwater recharge program like Orange County's, with even more community benefits. The model for this program starts by slapping a drinking water plant on the backside of one of Hampton Roads' existing sewage treatment plants so that the treated sewage that would normally end up in Chesapeake Bay would instead be purified by a drinking water plant. Once the water passes through the final stages of purification, it would then be injected underground (just as Orange County does). In addition to restoring the aquifer, one study suggests that the groundwater injections may be able to combat saltwater intrusion and prevent Naval Station Norfolk from sinking further.[7] One modeling study even predicted that SWIFT's groundwater recharge program—as it slowly replenishes the depleted aquifer—could *reverse* the land subsidence around the base, possibly raising the ground underneath the giant facility as much as three to six inches by midcentury.[8] In other words, SWIFT's groundwater recharge program just might be able to counter decades of land subsidence for the largest naval base in the world.

Then there are the benefits for Chesapeake Bay. By pumping millions of gallons of recycled water underground rather than discharging treated sewage into the bay, SWIFT is projected to reduce nutrient releases to the bay by 75 percent—reductions that the EPA was thrilled with as well.[9] The more Henifin's team looked at SWIFT, the better it sounded: the program would help the aquifer, help the Navy, and help Chesapeake Bay.

But what about the cost?

Henifin told me about the time his team first crunched the numbers. "They did this presentation to show me the results and they were kind of sheepish," he said. "They were like, 'Well, it's going to cost *a billion dollars.*'"

"And I'm like, 'Let's do it!'"

"And they were like, '*What?*'"

"It's a lot better than spending $2 billion on this darn overflow problem," Henifin said. "It's a much better investment, with much better outcomes for the environment, and it solves so many problems."

Let's check Henifin's math. The SWIFT aquifer recharge program was estimated to cost $1 billion. But federal officials were about to require $2 billion in water-quality upgrades—upgrades that the SWIFT program will take care of instead—hence the billion dollars in savings for his ratepayers. Henifin's SWIFT program shows that sometimes water recycling can be a very cost-effective water supply option, even in wet parts of the United States.

⌣

The program began with construction of the SWIFT Research Center in 2017—a pilot facility that has the ability to inject up to a million gallons of purified sewage underground per day. The center allows scientists and engineers to bench-test different technologies on a small scale, and the building also serves as an outreach facility where people can learn about SWIFT—and even taste the water. Three monitoring wells have been installed behind the research facility to track how quickly—or, rather, slowly—the injected recycled water travels underground. In 2022, after years of running tests in the Research Center, Hampton Roads Sanitation District began construction of the first full-scale SWIFT water recycling plant. Located in Newport News, the plant is slated to inject up to sixteen million gallons of recycled sewage into the aquifer daily by 2026.[10] Henifin's long-term vision is to add that same water recycling technology to four other sewage plants, eventually pumping one hundred million gallons of purified sewage into the aquifer daily by 2032. That would make it the largest potable water recycling program on the East Coast.

When the SWIFT program was in its formative stages, Henifin made the rounds, explaining its goals to public officials throughout the region in an effort to head off any potential opposition. There were lots of focus groups and open-house events. To Henifin, these public outreach efforts were time-consuming, but crucial. "I always—*always*—felt that, technically, we could make it happen, and regulatorily we could make it happen," he said, "but if we got the public skewed the wrong way on this, it was going to go the way of San Diego—and it would take another generation to come back around." Even though these outreach efforts all went well, Henifin found himself worrying about false alarms after the water tastings. He wasn't worried about his recycled water making someone sick, but he was worried about someone becoming sick from something else—like a bad lunch, spoiled leftovers at home, or whatever—and then falsely blaming his water tasting as the gastrointestinal culprit. "Anyone could have had an intestinal bug that day," he said. For forty-eight hours after every water tasting, he'd be on the lookout for a false alarm, but one never happened.

Despite Henifin's outreach efforts, questions arose about the SWIFT program. Remember that Orange County injects purified sewage underground and then it is treated again by drinking water utilities after extraction. San Diego also plans to treat its recycled sewage a second time, after it is withdrawn from a reservoir. What makes the SWIFT program unique is that thousands of homes and businesses in southeast Virginia withdraw groundwater directly from the aquifer via private wells, and this private well water does *not* receive a final round of treatment upon extraction. Even so, Henifin doesn't see that as a problem. "The rate at which the water moves through the ground is incredibly slow," he told me. "It's moving at something like thirty feet a year.... It could be a hundred years before our water molecules reach anybody's personal well nearby."

During that time, Henifin told me, the injected SWIFT water is continuing to be cleaned even more as it moves through the aquifer. "By then it's had plenty of time for nature to do its thing." What's more, he said, many southeast Virginians hold unique attitudes about their groundwater. "It's an interesting group," he told me. "For the most part, they don't want to know what's in their water, and they *definitely* don't want the government to tell them what's in their water—it's crazy." He described how the state extension service provides clinics for citizens to confidentially submit well water samples for testing. "So no one knows what's in your water except you," Henifin said. "They'd rather not know, or if they know, they don't want anyone else to know." What's more, most people in his service area don't drink groundwater. They're on surface water that is treated by regional municipal utilities that obtain their water from regional reservoirs. The result has been a community that, so far at least, has not been concerned about purified sewage being injected into the groundwater.

Regulation, or the lack thereof, proved to be a challenge. Like most states, Virginia had no regulations specifically targeting water recycling. The US government didn't either. Instead, water recycling is usually regulated under a combination of the Clean Water Act and the Safe Drinking Water Act. When the Clean Water Act was passed in 1972, the federal government generally delegated enforcement responsibility to the states, but Virginia officials took a pass on regulating groundwater injection projects. Thus the primary regulator for SWIFT—which, in the end, is a groundwater injection effort—was the EPA, not the state.

But the state *did* have regulatory authority over the sanitation district's sewage discharges, and when Virginia officials heard about the SWIFT program, they weren't quite sure what to make of it. "From the health department's perspective, this was a brand-new thing that had never been allowed before, and they were really concerned about the public perception," said Scott Kudlas, director of Virginia's Office of Water Supply. "There were just lots of unknowns," he told me. "[And] to do your first meaningful, managed-aquifer recharge with treated wastewater—at a scale like this—was kind of frightening and daunting from a regulator's perspective."

But the promise of SWIFT was alluring too. Norfolk wasn't the only place suffering from aquifer depletion. State studies showed that Virginia had been unsustainably mining its groundwater for years, and if the pace of water withdrawals continued it would create a three-thousand-square-mile cone of depression in the expansive Potomac Aquifer.[11] The situation had become so dire by 2009 that the Virginia Department of Environmental Quality determined that groundwater withdrawals had to be slashed by 50 percent. "We really didn't want to be in the position where we might close the aquifer down," Kudlas told me. "That's what we were panicked about." But SWIFT had the potential to wipe out thirty years of overpumping and replenish the aquifer throughout much of Virginia's coastal plain—even stretching into North Carolina. "It could restore the entire cone of depression," Kudlas said. Given that promise—and the lack of a public backlash—officials in the Department of Environmental Quality, the governor's office, and the Virginia legislature eventually embraced the SWIFT program. That was no small decision for government officials, who are rarely rewarded for taking chances. Kudlas said that the decision required a "willingness by everybody to accept each other's risks"—regulator and utility alike—as they embarked on the largest aquifer recharge program in state history, using purified sewage, no less.

Most water utilities tend to cringe at the thought of more regulation, but Henifin yearned for more state oversight, especially from health officials. "We saw, from the beginning, we had to have the Health Department involved," Henifin told me. "We couldn't have them say, 'Well, I don't know what they're doing.'" But when he approached health officials, he immediately ran into problems. They were uneasy about the groundwater injections but didn't have the resources to regulate them. "When we first learned about this project, we were concerned about its potential health impacts," admitted Marcia Degen, a Virginia health department official, "especially because of the large number of private drinking water wells in the area."[12] Degen and others tried to carve out time to review Henifin's project, but the workload was unsustainable. She wanted to hire someone to take on this new demanding role but didn't have the money in her budget. The legislature also refused to cover it. So Henifin cut a highly unusual deal: his utility would make a grant to the Health Department so that state officials could afford to hire someone to review his project. In other words, Henifin paid the state to regulate him—that's how desperate he was to ensure transparency and oversight for his innovative project.

Doesn't that create a conflict of interest? "It does," Henifin admitted, "but I didn't have any heartburn with us offering up the money for them, as a grant, for a full-time person to … ensure the public health is protected." So the sanitation district ponied up the money, and a Health Department staffer has shadowed the SWIFT program throughout development and implementation. Health officials said that Henifin's team has been "highly responsive in answering questions" and providing any data they have requested.[13] To round out the regulatory process, the state created an oversight committee and an independent monitoring laboratory to vet SWIFT's water samples. Henifin encouraged this move as well, as it could only help improve public confidence. But the state didn't want to pay for those oversight additions either. So, for the second time, his sanitation district paid the state for more regulatory oversight. "It's hard to say it's an independent monitoring lab when [the sanitation district] is paying for everything, right?" Henifin admitted. The state finally started picking up the tab for the independent monitoring lab and the oversight committee in 2022.[14]

~

As hard as it was for Henifin to patch together an oversight framework for his water recycling program, it would have been even more difficult if SWIFT was the first water reuse program in Virginia. But it wasn't. That distinction goes to the Upper Occoquan Service Authority (widely known as UOSA and pronounced "YOU-O-sa"). Located in northern Virginia, it is one of the oldest potable water recycling programs in the United States. It was founded in 1971, right before the Clean Water Act passed. Snaking over fourteen miles, Virginia's serpentine Occoquan Reservoir is a wide spot in the Occoquan River and forms the border between Fairfax County and Prince William County. Seated just west of the nation's capital, it is a key drinking water source for two million people.

This is DC commuter country, home to thousands of federal workers who keep the national government humming. What many of them don't realize is that UOSA dumps heavily treated sewage—from millions of people—directly into their drinking water reservoir. Normally, that treated sewage makes up about 8 percent of the Occoquan Reservoir's water, but during droughts, it can ramp up to 60 percent treated sewage.[15] In other words, at certain times a key source of water for the enlightened suburbanites of northern Virginia is heavily treated effluent, or what the industry refers to as indirect potable recycled water. What's more, these suburbanites have been drinking recycled water since the 1970s, making the Commonwealth of Virginia a quiet, no-drama leader in the potable reuse movement. Back in the 1970s, when Orange

County was building Water Factory 21, UOSA was building a potable reuse program in Virginia. One could think of UOSA's plant as the "Water Factory of the East," although it uses a different water recycling process than Orange County.[16]

How did UOSA become such a trailblazer? The Occoquan River first made headlines in the 1950s when published reports described one of its tributaries as a "slowly moving cesspool."[17] The watershed's sewage controversies became so prominent that they attracted the attention of President Dwight Eisenhower.[18] By the early 1960s eleven wastewater plants were dumping effluent into the Occoquan watershed—and these were pre-Clean-Water-Act discharges, with effluent that was just a step above raw sewage—or worse. "These plants were notoriously unreliable and experienced frequent failures and breakdowns," complained one report.[19] By 1963 the river was so maxed out with sewage that officials declared a moratorium on new plants.[20] At the time, 150,000 people downstream—including those living in Alexandria—got their drinking water from the Occoquan Reservoir.[21] By 1970 the reservoir was described as "seriously degraded," in "deep trouble," and on the brink of becoming a "sewage lagoon."[22] Complaints about algal blooms were rampant, as were taste and smell issues. Fish kills were common. As the suburbs around Washington, DC, continued to grow, an official report declared that pollution in the Occoquan "must be drastically curtailed" for the reservoir to continue serving as a drinking water source.[23]

These kinds of water quality controversies were occurring throughout the United States at the time. Public anger and disgust over rampant pollution in the nation's waterways would eventually lead to adoption of the federal Clean Water Act in 1972. As northern Virginia continued to grow, it became cornered by its own sewage and was mirroring the raging clean water debates that were echoing in the halls of Congress just a few miles up the road. Running out of options, people in northern Virginia started to wonder if they could treat their sewage to the point where it might be recycled as drinking water. "Should Northern Virginia residents drink their own refiltered sewage water?" asked a 1971 news story, "or should they channel it downstream to someone else?"[24]

Bob Angelotti has searing memories of the nation's water quality woes from that era. He spent part of his childhood in Ohio, and his dad took him on fishing trips to Lake Erie. "I can remember going there early on and catching boatfuls of fish," he told me, but in later years, "there were just dead fish everywhere." It's the kind of thing a boy never forgets. Then Angelotti's family moved to Washington, DC, "and it was kind of a continuation of the same saga with the Potomac River," he said. "We would get out there on the river, and below a certain point, you didn't go fishing [and] you wouldn't want to go wading."

Angelotti channeled those experiences into a career dedicated to clean water. Today he serves as executive director of UOSA, whose members include Fairfax County, Prince William County, and two local communities. The idea for UOSA was to band together on behalf of the Occoquan Reservoir before its crucial water supply became unusable. Officials decided to shutter all the unreliable treatment plants and transport that sewage to a single high-tech facility. The new plant would produce effluent that was near drinking water quality, which would then be discharged to the Occoquan River. From there it would flow twenty miles to a drinking water plant downstream—just as it always had—except that now the source water would be much, much cleaner than before.

The public loved it—except for the cost. In fact, there were so many water treatment protections and redundancies in the new UOSA facility that some critics called it overkill. But once the plant came online in 1978, the benefits became clear very quickly, and the cost concerns dissipated. "Water quality in the reservoir was like flipping a switch," Angelotti said. "You could immediately see huge improvements," which made the program even more popular. What about during times of drought, when the bulk of the water flowing into the reservoir is coming from UOSA's plant? That's not a negative. Angelotti's team considers UOSA's effluent to be the cleanest water flowing into the reservoir.[25]

Thanks to UOSA, water reuse has become normal in northern Virginia. Since 1978 the utility's production of recycled water has grown more than tenfold, from five million gallons to fifty-four million gallons per day. But Angelotti pointed out that water recycling is normal all over the United States—most people just don't realize it. That is especially true for cities located on rivers. Just about everyone on a river is downstream from someone else. Those upstream cities are pulling water from the river, using it, treating it to Clean Water Act standards, and discharging their treated sewage back into the same river, where the effluent mixes with the flow as it tumbles downstream. Then a few miles down, another city pulls drinking water from the same river— which is now a mixture of river water and treated sewage. On many rivers, this process is repeated again and again. In short, river people are almost always drinking the diluted, treated effluent from someone else upstream, but they just don't realize it. "Water recycling has been occurring in this country, and really all over the world, in an uncontrolled and unacknowledged manner—for centuries," Angelotti said. "It's unplanned. It's unregulated. It's de facto reuse."

That's exactly what the water industry calls it. The list of cities that practice de facto reuse is long and esteemed and includes Atlanta, Cincinnati, Houston, Nashville, New Orleans, Philadelphia, St. Louis, St. Paul, and Washington, DC. UOSA's wastewater treatment process is much more sophisticated than most, which means that de facto water reuse customers are often relying on a lower-quality source of water— not unsafe water, just not as rigorously treated. Ironically, the de facto water recycling facilities are the ones that somehow avoid the "yuck factor," whereas those that intentionally recycle wastewater—using more robust treatment technologies—are constantly worried about some public relations backlash.

Perhaps no one appreciates UOSA's program more than Jamie Bain Hedges, general manager of the Fairfax County Water Authority, which operates the drinking water plant twenty miles downstream. The Occoquan Reservoir supplies more than 40 percent of the water that Bain Hedges delivers to two million people every day. (The rest of her water comes from the Potomac River.) "About one in every four Virginians get their drinking water from us," she told me. Some of the nation's most important federal agencies headquartered in the commonwealth get their water from her utility. "The UOSA effluent is significantly better than what is coming off the rest of the watershed," she said. "It's a great story for northern Virginia, [but] it's something that we haven't really talked about a lot." The vast majority of her customers have no idea that they are drinking recycled water, she said. "Most of them don't understand how water gets to their tap, period."

The UOSA/Fairfax Water story represents one of the longest running, least controversial water reuse partnerships in the United States. Even so, there have been a few tense moments. One sore spot emerged a few years ago when UOSA considered selling some of its recycled sewage to a proposed power plant. Bain Hedges saw that as a threat to her source water. "We kind of raised a stink at the time," she said, "because, from our vantage point, the reason UOSA exists is to ensure the drinking water supply." Who would think that northern Virginians would be fighting over access to recycled sewage? The power plant proposal eventually fell through, but Bain Hedges fears a similar incident might arise in the future, so her point remains: UOSA's recycled water is so precious that it should be reserved solely for people to drink. "We've got to find a way to work with policy makers to either codify that in state code or in the Occoquan policy," she told me. "I don't think it's good public policy for it to be sold off and evaporated through some industrial process."

Like Orange County on the West Coast, UOSA's long, quiet, stable leadership in water reuse has been instrumental in helping other water recycling projects get off the ground. That includes Henifin's SWIFT program. "We're the beneficiaries of what was done at Occoquan," he said. "It was *huge*." Every time he gave a public talk about SWIFT, he told the audience that "a million people in Fairfax County are drinking this water and have been for forty years." UOSA was the first water reuse program in Virginia to set up an independent oversight committee, as well as an independent water quality lab, which is a key reason that the same state oversight system was implemented for SWIFT. "We just had a model to work from," Henifin said. "That was very, very important."

Henifin leaned on the example of Orange County as well, but that comparison was less direct. Why? Because Orange County uses a different process to recycle its water. It turns out that there's an East Coast method to water recycling and a West Coast method.[26] Orange County's West Coast method uses reverse osmosis to purify its water. The reverse osmosis system is more familiar to the general public but produces a brine that needs to be disposed of, which has environmental implications. It also takes a lot of energy to push the effluent through the reverse osmosis membranes, which is expensive. The East Coast method, used by SWIFT, Occoquan, and others, uses granular activated carbon, which employs substances like coal, coconut shells, and wood to remove contaminants. It can sometimes be cheaper because it uses less energy and there is no brine to deal with, but in certain cases the charcoal, wood, or coconut shells need to be disposed of in special landfills.[27]

Occoquan may have been the first water reuse program in Virginia, but from the beginning SWIFT has planned to be bigger—almost twice as large—at one hundred million gallons per day. In the process, Henifin hopes that SWIFT will help make the East Coast water recycling method even more popular. "I think it's going to show that you really

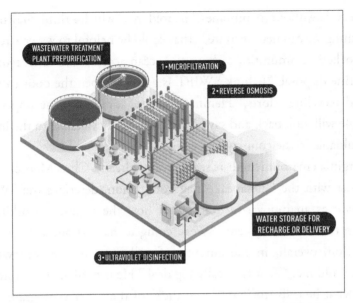

Figure 5-1a. An example of a reverse osmosis treatment system, which is informally known as the West Coast method of water recycling.

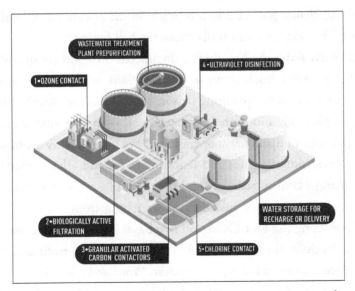

Figure 5-1b. An example of a granular activated carbon treatment system, informally known as the East Coast method of water recycling.

can recycle without membranes," he told me, "which I think may be the best thing we can demonstrate." That could be helpful to water-stressed, landlocked communities without the ocean access that's often required for brine disposal. "I think SWIFT is going to open the door to more inland recycling efforts," Henifin told me, adding that he hopes that "people will look back and say SWIFT was really valuable to the industry, valuable to the country as a whole."

Virginia environmentalists see the value already. Chris Moore, senior scientist with the Chesapeake Bay Foundation, describes the SWIFT program as an innovative example of how the region is working to restore the bay's ecosystem. "We're going to have about a 75 percent reduction, overall, in the amount of pollutants in the wastewater," Moore told me. "That is a really big deal." He is particularly impressed with what he calls the "can-do attitude" of the SWIFT program. "We can take something really challenging, like Chesapeake Bay, and we can find new and innovative ways to protect our ecosystem, and deal with additional challenges like a lack of water in the aquifer [and] land subsidence," he said. "We can really move the ball forward."

And what about the Navy? How do officials there feel about SWIFT? For that answer I reached out to Brian Ballard, who has a demanding job as the community planning liaison officer with the Navy's Mid-Atlantic Region. He monitors "encroachment"—any activities outside a base that might impact naval operations. "We have to be aware of what's going on outside the gates and fence lines," he said. That includes climate change, land subsidence, and flooding. His territory is expansive, covering thirteen major installations stretching from North Carolina to Maine along the East Coast and two bases in the Midwest. At Naval Station Norfolk, sea level rise and land subsidence are front and center for Ballard. As one of his superiors put it, "You can build your base to be Fort Knox, but if you can't get to work because of a Level Three Flood, you've kind of defeated the whole deal."

It's no surprise that enhancing resilience to climate change makes up roughly 25 percent of Ballard's job. "We deal with a whole range of encroachment-type issues," he said. "Climate has been one we've been engaged with for many years now—across the region—but particularly with Hampton Roads." As noted earlier, the sea is now eighteen inches higher at Hampton Roads—and Naval Station Norfolk—than in 1918. Eighteen inches is a lot for a base that sits just nine and a half feet above the Atlantic. That's why flooding has become such a problem both on the base and on many roads that lead to it. Because around nine of those inches are due to climate change—a number that is expected to grow in coming years—the ocean is Ballard's biggest concern. But the nine-inch drop due to land subsidence is not to be discounted. "Inches do matter," Ballard said, so the projections that SWIFT might just raise the ground underneath Naval Station Norfolk by up to six inches in coming decades could be a huge help. It won't solve the entire flooding issue, but Ballard said that if SWIFT can take land subsidence out of the equation, "it buys us time,… and the more time we have to deal with the problem, the better."

State officials are excited about SWIFT as well. As the head of Virginia's Office of Water Supply, Kudlas is responsible for sustainably managing the state's groundwater for generations to come. Earlier in his career he spent years working on Chesapeake Bay restoration efforts as well. Because SWIFT promises to provide so many benefits, Kudlas sees it as a transformative program that could have influence far beyond Virginia. "It's potentially one of the largest managed-aquifer recharge projects in the world, [and] it seeks to help solve a significant long-term water quality problem with the Chesapeake Bay," he told me. "I certainly think it would be a tremendous model for others to emulate." But 2032 is a long way off—that's when the SWIFT program is slated to be pumping one hundred million gallons of purified sewage underground daily. A lot can happen between now and then, which is why Kudlas can't help tempering his enthusiasm. "It's easy to get all excited and think of [SWIFT] as a silver bullet, but the reality is, it still has to prove itself," he told me. "While the preliminary returns are encouraging, we'll just have to wait and see."

⌣

That caution may be warranted. Henifin retired on February 22, 2022, and was replaced by Jay Bernas, a brilliant internal hire who was formerly director of finance. Bernas speaks highly of Henifin and is a big fan of SWIFT. "When we initially pitched [the SWIFT program]," Bernas told me, "people were like, 'Something's not right, because this almost sounds too good to be true.'" But two days after Henifin stepped down, Russia invaded Ukraine and global inflation skyrocketed, making major capital projects like SWIFT much more expensive than originally projected.[28] Just a few months after taking the reins, Bernas started to question whether he would be able to implement the grand water recycling vision that he had inherited. "That's really caused us to rethink this strategy and what our region can afford," Bernas told me. In our interview, just four months into his new job, Bernas openly wondered whether he could hit Henifin's goal of one hundred million gallons of recycled sewage by 2032.

Henifin's plan called for adding a drinking water facility to five of Hampton Roads' sewage plants. Bernas was now wondering whether adding just two drinking water plants—for a total of fifty million gallons of daily groundwater injection—would be enough. "What is the Goldilocks number for what needs to go into the ground?" he asked. "Is forty, or fifty million gallons … enough to stop the land subsidence?" Another option, he said, was to disperse the cost of the SWIFT program over a longer period of time, which would make the project more affordable—like using a mortgage to buy a house. "I think eventually it will all get done," he said, "[but] it becomes more affordable if you can spread things out."

Bernas was in a tough spot. Just starting out in a demanding new role—trailing in the wake of a charismatic, visionary leader—and right out of the gate, he faced a difficult decision. It was clearly a struggle for the former finance director not to worry about the money. But it also felt like the vision had left the building when the visionary retired. With Henifin gone, the innovative, award-winning SWIFT program seemed to be backsliding. In one of my final interviews with Henifin, we talked about how his legacy may be morphing. Comfortably retired and detached from day-to-day decision making at Hampton Roads, he had no idea. Sometimes journalists have to be the bearers of bad news, and that was one of those moments. He was clearly disappointed.

But the news reignited the passion in the proselytizer. "I'm not overly surprised," Henifin told me. "We've got a new general manager—maybe a little risk-averse to begin with—not wanting to push the envelope with cost increases. But man, it's hard to imagine any better investment for the long-term—all the benefits that SWIFT provides—and it was never just about putting in enough [water] to slow down the land subsidence. This is the right thing to do for wastewater—period! We should not be putting it back in the environment unless it's in the right condition. And if it can be reused to some benefit by future generations, by all means we should be doing that—whether you are at [Hampton Roads] or some other place. I think they're missing the bigger point.... I don't even know what to say. It's kind of sad."

I asked if the news gave him retirement remorse.

"No, not really," he answered. "I mean, I had to go at some point, and I wasn't going to hang around until 2032 when it was all built out."

And what about cutting SWIFT in half—to fifty million gallons per day? Would that be enough?

"I don't think it's enough. I think we should be going for zero discharge at all our plants," he said. "We ought to figure out how to make that happen—lead the way! That's how it should be everywhere in the country.... If we don't get every drop in the ground here at Hampton Roads, maybe someone else will."

SWIFT has already been transformative. Henifin has changed the way the Commonwealth of Virginia thinks about sewage. The only question now seems to be whether SWIFT's full potential will be realized—or if a portion of the original plan will have to do. Henifin's vision has captivated the national water reuse community as well, with one expert dubbing his program the "win-win of win-wins."[29] The first SWIFT plant is under construction and is scheduled to inject sixteen million gallons of sewage into the aquifer per day by 2026. Once that plant is up and running, valuable lessons will be learned—for Hampton Roads, the Commonwealth of Virginia, the EPA, and the national water reuse community. Perhaps the biggest lessons will be about risks and trade-offs—financial and otherwise—and about the courage it takes to let go of the old way of doing things and fully step into the brave new world of future water.

CHAPTER 6

Running Dry (Almost) in Texas

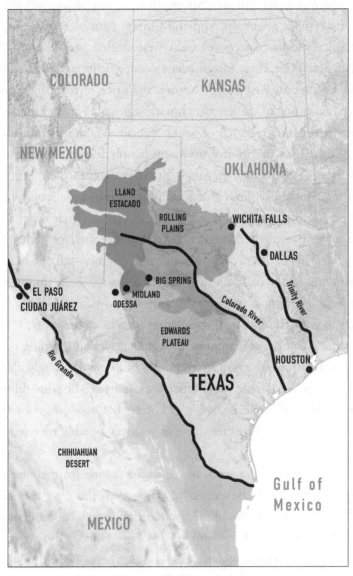

Texas

WHEN THE PLANE TOUCHED DOWN AT MIDLAND AIRPORT, I was amazed at how little the West Texas steppe had changed since my last visit to this arid landscape decades before. Pumpjacks and oil rig flares were visible all the way to the horizon. Sitting atop the petroleum-rich Permian Basin, everything about the Midland-Odessa Petroplex cries "oil!"—as it has since the first successful wells were drilled here in the 1920s.[1] This is where West Texas Intermediate crude got its name. Despite the region's boom-and-bust energy history, the price of that "light, sweet" crude remains one of the key benchmarks for global energy markets.

But I had come in search of water, not oil. My destination was Big Spring, a town of twenty-five thousand people forty miles to the east. In 2011, as Texas battled the single driest year on record,[2] Big Spring garnered national headlines after local officials asked the state for emergency permission to add purified sewage directly to their water supply. It was an unprecedented moment in US water management history. Unlike San Diego, Orange County, or even Virginia—all of which took the extra precaution of "buffering" recycled sewage in an aquifer or reservoir before adding it to their drinking water—officials in Big Spring wanted to skip the buffer. Instead, they sought permission to implement what is known as direct potable water recycling—adding purified sewage straight into the water system. No buffer. Surprisingly, the state agreed, giving West Texas water officials clearance to do something that had never been allowed in the United States before—except by NASA (orbiting astronauts have been drinking direct potable recycled water for years).

It was a big step for a small town, and everyone from congressmen to third graders suddenly wanted to visit—even the *New York Times*.[3] During the flurry of national media attention, one local resident joked that he looked forward to drinking his beer twice.[4] The water recycling world knew that one day a city—somewhere—would be the first intrepid community in the United States to add purified sewage directly

to its water supply. But what was surprising is that it turned out to be a
rush job, in a small town, out on the oil patch, during a drought emergency—in a state that is not exactly known for robust regulatory oversight. Given the hypersensitivity of the water reuse community, many in
the industry did not consider a small town in West Texas to be the ideal
venue for a direct potable reuse debutante ball, but sometimes history
has other plans. Big Spring's precedent helped foster two other similar water recycling projects in the state, making Texas the unexpected
national leader in the most complicated, controversial, and cutting-edge
form of water recycling.

⌐

Big Spring sits in a unique place on the Lone Star landscape. It's where
three natural regions converge: the Rolling Plains tumble down from
the north and east, the rangelands of the Edwards Plateau rise up from
the south, and the High Plains, or Llano Estacado, extend down from
the panhandle.[5] In 1768, when Spanish explorers traversed this dry ter-
rain, they found Comanche and Pawnee tribes dueling over the spring's
precious waters.[6] Settlers and railroads tapped the spring as well, and
by 1925—thanks to overpumping and development—the spring dried
up. Today it is among a handful of springs in the state that are "maintained" by pumping water into them.[7] The town of Big Spring now gets
its water from the Colorado River—no, not the Colorado River that's
home to Hoover Dam, but the *other* Colorado River, that starts on the
Llano Estacado and cuts through the heart of Austin on its way to the
Gulf of Mexico. Three reservoirs on the river serve as the primary water
source for Big Spring, as well as Midland, Odessa, and other towns in
the region.[8]

But during the historic drought of 2010–2015, two of those reservoirs withered into glorified mud flats, and the third was hovering at
just 12 percent full.[9] That plunged the Colorado River Municipal Water

District into a full-blown crisis as it scrambled to deliver more than fifty million gallons of water each day to six hundred thousand West Texans.[10] As one local water official put it, "It's not a case of lakes drying up—it's *all* the lakes drying up."[11] The stakes were unusually high given the role that the region's economy plays in global oil markets. But it was also a reminder that although oil pays the bills, water is the foundation of the Petroplex economy, just like it is everywhere else. "I worked seven days a week for two years," said John Grant, the soft-spoken native Texan who led the water district through the drought. "I sleep now," he told me. But back then? "No, you didn't sleep."

Fortunately, Grant had long been considering a direct potable reuse project. The water district had commissioned feasibility studies in the early 2000s and a pilot project in 2009. Development of his nascent water recycling operation shifted into high gear during the drought. But Grant quickly ran into headaches with the Texas Commission on Environmental Quality (TCEQ). The state didn't have regulations for how to build a direct potable reuse plant. "I gritted my teeth," Grant said, "but the TCEQ did not have any rules for what we were doing." So he worked closely with state officials—during a drought emergency—to literally make it up as they went along. The water district and the state eventually came to agreement on rules "that were reasonable, that we could live with," Grant said, "[but] it took a while."

The construction contract for the historic project was awarded in May 2011, and the plant was recycling water just two years later. The $14 million facility captured most of Big Spring's effluent, treating it with microfiltration, reverse osmosis, hydrogen peroxide, and ultraviolet oxidation. The purified water was pumped into a large pipe that was already distributing reservoir water to the various communities in the district. Once the blended water arrived at a town, it passed through a conventional drinking water treatment plant before being distributed to homes and businesses.

As usual, reverse osmosis produced highly concentrated briny waste-water that had to be dumped somewhere. Most reverse osmosis plants are on the coast and discharge their brine to the ocean, but Big Spring is more than four hundred miles away from the Gulf of Mexico. Not to worry, Grant told me. The Permian Basin is shrouded in salt, which means that much of the regional groundwater—and many surface waters—have very high salinity rates. So the Big Spring plant discharges its brine straight into Beals Creek, which—remarkably—happens to have a higher salt content than the water district's concentrated brine discharge. "We're improving the quality of the water in the creek," Grant quips. Not quite. The brine includes more than just salt. It also contains contaminants that have been extracted from Big Spring's effluent as it is transformed into drinking water.

Given the new water recycling plant's national significance, West Texas media outlets followed the facility's construction with blow-by-blow reporting. Despite the plant's unprecedented configuration, the local coverage generated no notable controversy. The desperation of the drought certainly played a role in this muted public reaction—it's hard to be picky about your water when it's about to run out. But there was another factor that I learned during my visit to Big Spring: most people there don't drink water straight from the tap. Grant doesn't, and he ran the water district for more than twenty-five years. Neither does the district's operations manager.[12] A former Big Spring mayor I interviewed avoids it too.[13] This situation has been going on for years—long before the new water recycling plant came online. Instead, many people have small reverse osmosis systems in their homes, usually at the kitchen sink. According to one local vendor, perhaps 75 percent of the local population uses in-home water treatment for drinking.[14] Many of the rest drink bottled water—reserving the tap for bathing, laundry, and flushing toilets. Big Spring's water has long harbored a bad reputation,[15] but it's not just Big Spring. Remember that all the communities in the region rely on the same source water, the bulk of which comes from reservoirs with high salinity rates.[16]

I discovered that the hard way. I'm a proud tap water guy—whether it be recycled or otherwise—because it's the most sustainable water you can get. It's also the cheapest. When I arrived at my hotel in Odessa, I did what I always do and curiously sampled the local tap. But for the first time I can remember, I struggled to choke it down. The taste was a repugnant mix of salt, minerals, and metal. To be clear, I'm no water snob. During wilderness vacations I regularly use a backpacking filter to gather water from all manner of lakes, ponds, and mudholes. But this water was something else. "It's kinda salty," Grant admitted, "but it's the best we got." No wonder the new plant was devoid of controversy, I thought. When people don't drink the water, it's hard to get too worked up about the first direct potable reuse plant in the United States adding recycled sewage to your water supply.

Grant said that people *did* get worked up after the national media came to town. He said that the out-of-state reporting was saturated with hype and sensationalism. "West Texans were fine with [the new facility]," he told me, "but as soon as the plant went online, CNN and [others] came out here, and it went national, and all bets were off.... They put a whole different spin on the story," he complained. "'Toilet to tap,' 'people are drinking their wastewater,' yadda, yadda, yadda."

Sensationalism aside, Big Spring's new plant was a historic moment in the water recycling movement. What makes direct potable reuse so newsworthy is that there is very little margin for error, a nuance that most journalists—local or otherwise—failed to recognize in their coverage. Because there is no environmental buffer, in the rare event that a dangerous contaminant slipped through the treatment process, officials would need to detect it immediately and then divert the contaminated water offline. Such a diversion system is standard at all water recycling facilities, including Big Spring's. The town's plant has real-time monitoring devices to detect contaminants—but not as many as I thought it would, especially given that they were making history with direct potable reuse.

Remember the acetone incident in Orange County? When an industrial customer illegally dumped a large amount of the organic solvent into the sewer system? Acetone passed through Orange County's wastewater treatment plant undetected, only to set off alarms in the water recycling plant next door. As it happens, acetone is one of the few contaminants that is *not* screened out by reverse osmosis. By the time Orange County determined that the contaminant was acetone, the solvent had passed through the water recycling treatment system and had been discharged into the aquifer. There it became diluted and dissipated and was not detected again, despite regular groundwater monitoring.[17] California officials said that the acetone never got close to entering the drinking water system.[18] That's the beauty of having an environmental buffer, Mike Marcus, Orange County's general manager, told me. Just in case something nasty slips through, it doesn't go straight into the drinking water supply. Instead it ends up in the buffer, giving officials more reaction time.

Big Spring doesn't have a buffer, and it depends on reverse osmosis just like Orange County, which means that acetone could slip through there too. Grant had not heard of Orange County's acetone incident until I told him about it. Neither had the district's operations manager, John Womack. So in an interview with both of them, I asked if they do real-time monitoring of total organic carbon, like Orange County does, to detect acetone?

"No, we don't," Womack said.

"So, what you're saying," Grant added, "is they have a way to detect it, and we don't?"

"Right," I responded.

This lack of monitoring is not a hypothetical issue. While I was in

Big Spring the city manager told me about a time when an industrial truck inadvertently dumped so much acetone into the city's wastewater that it temporarily disrupted operations.[19] I mentioned Big Spring's acetone incident to Grant and Womack, and for the next several minutes we talked about acetone, real-time monitoring, and concerns that people have about the lack of an environmental buffer with direct potable reuse. We also talked about concerns that people in other parts of the United States have with Texas being the first to experiment with direct potable reuse given the state's reputation for having a lighter regulatory hand than, say, California. Then, finally, I asked if it might be worthwhile for Big Spring water officials to consider adding real-time monitoring for acetone to their history-making plant?

"It may be," Womack replied.

⌣

Although the Big Spring plant made water recycling history, from the beginning Grant knew that it would not be enough. The new plant produced fewer than two million gallons of recycled water per day. Compare that to the water district's daily demand, which fluctuates between forty-two million gallons in the winter and seventy-two million gallons in the summer.[20] So, even though the new water recycling plant was helpful, it was supplying just a fraction of the overall daily need. Even with the new facility coming online, Grant was desperate to find additional water to augment his desiccated reservoirs. It was a two-pronged emergency approach: (1) quickly build the nation's first direct potable reuse plant (2) while also combing the landscape for new water supplies. Eventually Grant tracked down groundwater in a neighboring county. That meant that while he was rushing to complete his precedent-setting direct potable reuse facility in 2013, he was also frantically laying sixty-five miles of pipeline to access up to thirty million gallons of groundwater daily.[21] No wonder he didn't sleep. Ironically, although his direct potable reuse plant drew national headlines, it was the pipeline that actually saved West Texas from the strangling drought.

⌣

Then, a year later, it all happened again—same drought, same state, same problem—different city. More than 100,000 people in Wichita Falls were running out of water. Unlike Big Spring, Wichita Falls did *not* have a viable groundwater option, and by early 2012 officials estimated that there were just eighteen months of water left in the local reservoir system. The city, 150 miles northwest of Dallas, declared a "Stage 5" drought, the most severe level, and "Pray for Rain" signs popped up all over the area. Local conditions were drier than during the Dust Bowl period of the 1930s.[22] People started to panic. Daniel Nix, the unflappable civil servant overseeing the city's utility operations, told me about distressed residents calling him with all sorts of angst-ridden questions. "'Do I need to sell my business?' 'Do I need to sell my house?' 'How am I going to sell my house?' 'Who's going to buy a house in a town that's running out of water?'"

With no other options, Nix followed Big Spring's lead. He approached the state for permission to build an emergency direct potable reuse plant that was more than twice the size of the West Texas operation. Wichita Falls wanted to use the most controversial form of recycled water to supply half the city's need of ten million gallons per day. After some hesitation, the state gave Nix the go-ahead to write up a concept paper on how he planned to develop the largest direct potable reuse program the United States had ever seen. But, for the second time, Nix said, state officials were making it up as they went along. "There were no rules. There were no standards. There was nothing—except for Big Spring," Nix told me. So like Grant before him, Nix spent months wrangling with state officials over a long list of potential parameters that would end up dictating what his direct potable reuse operation would look like. And it was all rushed, due to the drought. "From the time we approached them … to the time of turning it on, it was twenty-seven months—with no regulations and no guidance," Nix said. "That's breakneck speed."

State officials were helpful, but nervous. Nix said that they were concerned about getting sued if the project went sour. He and his boss were nervous too—there was scant guidance in the literature about how to build a direct potable reuse plant. There was plenty of literature on how to set up an *indirect* potable program (like those in Orange County or Virginia), but not for the facility Wichita Falls wanted to build. The pressure was immense. "If this didn't work," Nix remembered thinking, "obviously we were going to be litigated. We weren't going to have a job, [and] nobody would touch us for a job with a ten-foot cattle prod." Nix dove into the research, but the internet provided limited help, so he spent countless hours poring over books and journal articles obtained through interlibrary loans. He became intimately familiar with Big Spring's facility and tried to find out as much as he could about the world's only other direct potable reuse operation, which was eighty-five hundred miles away in Windhoek, Namibia. But there wasn't a lot of detailed information available about that program either. "There were times," Nix told me, "when you're sitting there going, 'God, I hope I get all of this right.'"

Wichita Falls' dilemma was different than Windhoek's. It was different than Big Spring's too. Nix already had a reverse osmosis plant that was desalinating water from a salty local reservoir. (The plant used microfiltration as well.) But during the drought, the reservoir shrank and salinity concentrations skyrocketed, maxing out the capabilities of the desalination facility. The reservoir water had become too salty to use, and the city was about to shut down the plant.

Nix proposed keeping it open to recycle wastewater instead. The idea was to pipe effluent from his sewage treatment facility over to the soon-to-be-idled desalination plant, where the plant's microfiltration system and reverse osmosis membranes could be used to purify wastewater rather than saltwater. Then the purified effluent could be added to the drinking water supply. Most importantly, Nix's plan was temporary. He only wanted to use direct potable reuse to get through the drought. Unlike Big Spring, he was not interested in having direct potable reuse become a permanent part of his water system.

The state finally signed off on Nix's plan—in principle. But officials understandably demanded months of testing before the project went online. Then they wanted to see things for themselves before formally approving the operation. Once again, Nix was struck by how much everyone was winging it. "They sent their high-level people," he recalled. "None of them had ever operated a water plant, and none of them had ever been to a wastewater plant." He couldn't believe it. "We were teaching the regulatory body how to do this, and they were regulating *us*." After spending two weeks on site, the state finally signed off on Nix's plan in June 2014,[23] and twelve days later, he was adding direct potable reuse to Wichita Falls' drinking water.

It was another historic moment for the water reuse movement. After decades of controversy, jokes, and doubt about *indirect* potable water recycling—which took the extra step of "buffering" purified sewage in an aquifer or reservoir—two Texas cities, in two years, started adding purified sewage directly to their water systems, without a buffer. It should have created a stir, but it didn't. "The public support was overwhelming," Nix said. People were concerned—not about the quality of their water, but about running out of it. The level to which Wichita Falls residents embraced direct potable reuse was extraordinary, especially when compared to the revulsion that swept through San Diego just fifteen years before. Wichita Falls residents printed T-shirts that proudly announced, "We put the No. 2 in H2O!" Another shirt proclaimed, "Wee Recycle!"[24] The community responded to the drought with major conservation efforts as well, which cut daily domestic use by 65 percent.

They conserved so much that Nix started to worry about having enough sewage to recycle.

San Diego's original controversy showed that proposing a potable water reuse program in the absence of a crisis can backfire. But Big Spring and Wichita Falls showed that if you wait until it's almost too late, people can be extremely supportive. The national media gave some coverage to Wichita Falls' new program,[25] but not at the level that would have occurred if the water strife had struck a major metro area. But the new plant couldn't escape the late-night comedy circuit. In 2014, Jimmy Fallon's copywriters tucked the following joke into his *Tonight Show* monologue: "A town in Texas just announced a controversial plan to recycle toilet water and use it for drinking water," Fallon joked. "[Then the] dog said, 'How are you only thinking of this now?'"[26]

Wichita Falls was unfazed by the ribbing. Award-winning local media coverage was crucial in building the community's self-confidence.[27] Nix worked hard at being transparent and at forging relationships with local reporters. He invited journalists to regular brunch briefings to emphasize just how desperate the drought had become—and how crucial it was for the community to accept potable water recycling as a solution. "*We need your help* to educate the citizens on this," he begged. One local television station devoted its entire thirty-minute broadcast to the drought. The *Wichita Falls Times-Record-News* published a comprehensive series of stories called "Lifeline: Covering Every Drop"[28] that Nix said played an important role in keeping local citizens informed. "They were *really* helping," Nix said of the local media. National and regional coverage was less reliable, however. One *Fort Worth Star-Telegram* article described Wichita Falls residents as drinking "potty water."[29] But even then, local journalists tried to put the mocking publicity in perspective. "Generally speaking," one local report said, "all of the stories focus on how gross the concept seems at first, before they explain the science behind it, and that it actually might make sense."[30]

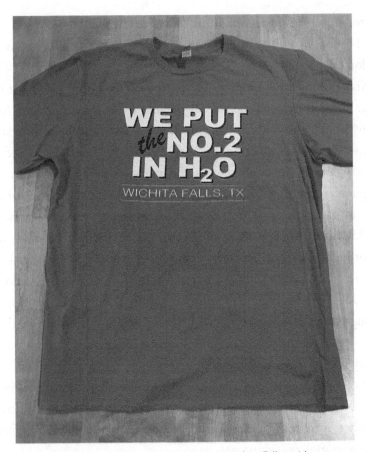

Figure 6-1. During the Texas drought of 2010–2015, Wichita Falls residents enthusiastically embraced their emergency direct potable water recycling program, even printing jocular T-shirts that supported the city's water reuse initiative.

Public support was not universal. Many locals harbored doubts about direct potable reuse, yet those qualms were never strong enough to foment any organized opposition. Many concerned locals switched to bottled water—sales of which rose by 9 percent after the new plant came online.[31] But Nix saw that as a way to augment his water supply further. "That's fine if you're going to drink bottled water," he told me with a smirk. After people drank the water and flushed their urine down the toilet, he would recycle it for others to drink. "I'm going to reuse that bottled water," he said, "so go ahead.... You're helping me out in the long run."

Nix's groundbreaking program ran for twelve months without incident. He is convinced that direct potable reuse saved his city. "Absolutely, it did," he said. But during the last few months of operation, the project became superfluous. Why? Because it rained and rained and rained again. In a matter of weeks, the reservoirs were full. Wichita Falls is a pious community, and many locals saw it as a miracle—their prayers for rain had been answered. From the beginning, Nix had planned to flip his direct potable reuse program into an indirect program once the drought was over. The idea was to pump his recycled sewage into a local reservoir before adding it to the drinking water system, leaning on the added security of an environmental buffer. But when news spread through town that his *direct* potable reuse operation was shutting down, the true popularity of the program became apparent. "That's when the outcry started," Nix told me.

Local residents had two main concerns. First, they found the drought to be so emotionally scarring that they worried that a dry spell might quickly return and the city would be unprepared if the direct potable reuse operation was shuttered.

And the second reason people wanted to keep the program?

"They liked the taste."

CHAPTER 7
El Paso's Quiet Leadership

El Paso and Ciudad Juárez

BIG SPRING AND WICHITA FALLS WERE THE EARLY ADOPTERS, but no community in Texas has embraced direct potable water reuse more bullishly than El Paso. Venerable and diverse, the city is surrounded by the chalky Chihuahuan Ecoregion, the largest desert in North America.[1] Throughout El Paso's history, water innovation has been key to its survival. In many ways, the city has been an unsung leader in the water recycling movement—its influence often overshadowed by Orange County, California, or Occoquan, Virginia. In the late 1970s when Occoquan and Orange County were heralded for embarking on some of the country's first potable water reuse projects, El Paso was not far behind. But somehow its water recycling leadership does not receive the same level of attention.

Perched on the bank of the Rio Grande, this border city has long relied on two key water sources: the local Hueco Bolson Aquifer and the river. Throughout much of the twentieth century, the aquifer was king, but in 1952 the US Geological Survey called out El Paso's unsustainable groundwater pumping, showing withdrawals exceeded the natural recharge rate by five million gallons per day.[2] A decade later a consultant reported that the overpumping was getting worse,[3] and studies repeatedly predicted that the aquifer would run out of water in a few decades.[4] Officials tapped a second, smaller aquifer, but more needed to be done. "In the long run we're looking at drinking recycled wastewater," one local official predicted in 1976. "First, we'll use recycled water for industrial purposes, then later on for recharging our underground wells, and much later … for drinking water."[5]

That prediction proved prescient. In 1978 the city adopted plans to build an indirect potable reuse groundwater recharge plant, and by 1985 the $33 million facility opened with a capacity to produce ten million gallons of recycled water daily.[6] Like Occoquan in Virginia, El Paso officials went with the East Coast method of water treatment, purifying the city's effluent with granular activated carbon, which uses materials such as coconut shells or coal to remove contaminants from the water.[7] The new recycled water was pumped into the Hueco Bolson aquifer, where it would spend years underground before being withdrawn and treated again for drinking. That made El Paso a leader in potable reuse, but the city's water situation remained tenuous. The new plant did not come close to making up for the groundwater pumping deficit.

There was another problem: the city was not entirely in control of its own destiny. El Paso's water fate has always been wedded to Ciudad Juárez, across the river in Mexico. With nearly seven hundred thousand people, El Paso is the sixth largest city in Texas. But when you add in another one and a half million people in Juárez, it's easy to see why the combined metro area has been called the largest bilingual, binational workforce in the Western Hemisphere.[8] All those people share the same declining aquifer. In the 1990s, under the leadership of Ed Archuleta, El Paso's esteemed water manager, the city imposed unpopular but integral water conservation policies.[9] Archuleta, who retired in 2013, also embarked on a long-term effort to shift much of the city's water consumption away from the declining aquifer and over to the river. "The Rio Grande, long ignored, and often abused," one news story said, "has become the city's hope for the future."[10] By 1996 El Paso was getting 50 percent of its water from the river, up from just 20 percent a few years before, which brought relief to the groundwater system.[11] But on the other side of the river, Juárez was still pulling 100 percent of its water from the aquifer, which continued to decline.[12] As water options dwindled, economic implications mounted. El Paso soon started turning away prospective water-intensive industries, including circuit board manufacturers, metal platers, and stone-washed jean factories.[13]

So El Paso innovated again. In 2007 it built the largest inland desalination plant in the United States. It was an $87 million partnership with Fort Bliss, the local Army base, which shared El Paso's water worries.[14] The aquifer under El Paso has different layers of water, with rich freshwater near the top and undrinkable brackish water as you move farther down. The desalination plant tapped into these salty dregs to produce up to twenty-seven million gallons of water daily—15 percent of El Paso's peak demand.[15] Under an unusual arrangement, the state allowed El Paso to inject the desalination plant's waste brine into yet another salty aquifer more than four thousand feet below ground and twenty-two miles away from the city.[16] The desalination plant was an extremely important boost to El Paso's declining supplies, allowing Fort Bliss to grow during a time of expanded desert warfare around the world.

But then the Rio Grande started to decline as well. By 2022 the river had been reduced to a sliver as it trickled into El Paso. Today the riverbed is dry, or nearly dry, most of the year, other than during a few months (or mere weeks) in the spring when flows temporarily increase. The spindly river created a conundrum for city leaders: How do you provide for seven hundred thousand people when both of your major water sources are in decline? *And* you've already built an indirect potable reuse plant, *as well as* a desalination plant?

You innovate again—by building the nation's largest direct potable water recycling facility. It's a fascinating water tale. From the 1940s to the 1980s, El Paso leaned heavily on a declining aquifer system and built the state's first indirect potable water reuse facility. Then the city spent the next few decades boosting water conservation while also pivoting away from the aquifer to the river *and* constructing the nation's largest inland desalination plant. But as the Rio Grande shriveled into the Rio Pequeño, the city's innovative leaders pivoted again, this time announcing plans to build the nation's largest direct potable reuse facility. "Diversification is the goal, rather than desperation," said Christina Montoya-Halter at El Paso Water. Big Spring and Wichita Falls paved the way, but El Paso was making unprecedented investments in direct potable reuse. El Paso's new Advanced Water Purification Facility will be twice as large as the temporary direct potable reuse plant in Wichita Falls and five times larger than Big Spring's. The plant is scheduled to start recycling ten million gallons per day in 2026. It's a drought-proof water supply that will keep on giving. "As the population increases," the water utility's website says subtly, "there will be more treated water to purify."

Gilbert Trejo is responsible for building the new plant. Spirited and gracious, Trejo is an effusive champion of water reuse and a past president of the national WateReuse Association. As El Paso Water's vice president of operations and technical services, he sees the new plant as the logical next step for his desert community. "The day's going to come when there's no river water," Trejo told me. "We need another reliable, drought-proof, local water supply and the only way to do that is by recycling our own wastewater for drinking."

Trejo grew up in El Paso, and his first job out of high school was cleaning the city's numerous irrigation canals. After getting two engineering degrees, he spent several years building wastewater—and drinking water—plants, which is the perfect background for a career in water reuse. Eventually he ended up at El Paso Water just as the utility's cerebral president, John Balliew, was contemplating a new direct potable reuse facility. The city already had a successful *indirect* potable reuse program that was recharging the aquifer, so officials considered expanding that effort first. But the logistics of that idea, which included building a large and costly pipeline right through the congested city, didn't pencil out. A direct potable reuse plant was cheaper and less disruptive to the community.

As the city had long been recycling water via an environmental buffer, the public was unconcerned by the thought of adding direct potable reuse. Polling showed that more than 80 percent of the population supported the program.[17] Fortunately for Trejo, the state's regulatory scene became more clarified as well. He told me that the Texas Commission on Environmental Quality (TCEQ) had a "framework" for vetting El Paso's project, leaving him with a more functional regulatory experience than Wichita Falls or Big Spring had encountered. "I found them all to be very thoughtful.... They had a lot of very good questions," he said of the TCEQ officials. "They put some stringent requirements on us, and rightfully so."

As the state's third direct potable reuse facility moved into the planning stages, pressure mounted on the TCEQ to come up with some sort of regulatory guidance. Daniel Nix was so scarred by the lack of direct potable protocols during Wichita Falls' drought emergency that he spent years afterward lobbying for more concrete direct potable reuse rules. "We didn't ask for regulations," Nix told me. "We just asked for a guidance document." In 2021 Texas passed legislation requiring the TCEQ to create "regulatory guidelines" for direct potable reuse. Texans call those rules RGs for short. What that means, Nix said, is that the next time a city wants to build a new direct potable reuse plant, the state "can hand them this RG and say, 'Alright, here's the process that you're going to go through'"—something he wished he had in 2014.

Although the new guidelines are an improvement, they are nothing compared to the California approach. For years, officials there worked tirelessly to create detailed, science-based regulations for direct potable reuse. While Texas was approving plants on an ad hoc basis, California was still drafting and redrafting policies and procedures, eventually releasing final direct potable regulations in 2023—more than ten years after Texas had signed off on the Big Spring plant. It's a classic situation, with California having a reputation for being one of the most heavily regulated states in the country and Texas much less so.

That dichotomy has made the "all-for-one and one-for-all" national water recycling community uneasy as people diplomatically try to straddle the classic Texas/California regulatory divide. The fear, especially outside Texas, is that the state's more flexible regulatory environment increases the risk of an "incident" happening—like a plant failing to provide safe, clean recycled water—possibly setting back the entire industry

Water Reuse Cost Comparisons (per acre foot)

	El Paso	Wichita Falls	Orange County	San Diego
Groundwater	$200	N/A	$624	N/A
Surface water	$300	$325	N/A	N/A
Desalination	$600[a]	$975[b]	N/A	$2,975[c]
Direct potable	$1,000	$684	N/A	N/A
Indirect potable	$1,400	$902	$850	$1,900[d]
Imported water	$3,000	N/A	$1,200	$1,100

Notes: There are 325,851 gallons in an acre foot of water. N/A = not applicable.
[a] Brackish groundwater.
[b] Brackish surface water.
[c] Ocean water.
[d] Estimate.

Table 7-1. How the cost of recycled water compares to other sources of supply varies widely depending on location. (Sources: El Paso Water, City of Wichita Falls, Orange County Water District, City of San Diego, San Diego County Water Authority)

for years. "Yeah, it *is* a concern—absolutely," said Mike Markus, the highly regarded general manager of the Orange County Water District. He described Texas's practice of approving direct potable reuse plants without regulations as "really rolling the dice." Markus spoke highly of Gilbert Trejo and El Paso Water, and he was not concerned about their direct potable project, but he had broader concerns about the Texas approach, particularly with regard to smaller communities with limited resources. "I'm sure [El Paso Water] will do the right thing," he told me. "They're a good agency. But you don't have agencies like that all around—particularly in Texas.... All it takes is one misstep and it sets the entire industry back. So, that's a huge concern."

⤙

Trejo has heard the out-of-state rumblings but defends the Texas approach. "It's not a simple process," he said. "What we have works for us—works for Texas anyway." He told me he fears that the more restrictive California regulations will make it too difficult and too expensive for many water-needy communities to adopt potable recycled water. What's more, he said, each wastewater plant has different purification needs depending on the type of sewage that is being treated (industrial, residential, or a combination), which is why he appreciates how the Texas approach allows regulators and utilities to customize each new water recycling project for a given community. "That's why I think the Texas framework is really good, because it's not cookie-cutter," Trejo said, adding that California's "prescribed" regulatory approach "can almost hurt as much as it helps."

Officials at the TCEQ declined to be interviewed for this book, but they agreed to answer a series of written questions. "The use of wastewater effluent as a source for drinking water is reviewed on a case-by-case basis," officials wrote in an email. "...Each [direct potable reuse] project is unique and requires an evaluation process that takes site-specific characteristics into account to be protective of public health." Officials added that this regulatory approach "avoids unnecessary 'overdesign' of the plant."

But Trejo feels the pressure. That's why he's building a water reuse plant featuring extraordinary safety precautions and redundancies—redundancies that tellingly go above and beyond what state regulators require. El Paso Water has already installed real-time monitoring devices throughout much of its sewer system—one of the first public utilities in the United States to do so. That should give El Paso Water advanced notice of dangerous contaminants that end up in the wastewater, long before they reach the sewage treatment plant, not to mention the water purification facility downstream from the sewage treatment plant. There will be extensive real-time monitoring inside the new direct potable reuse plant as well, and if the alarms go off, the problematic water will be immediately diverted offline.

In a highly unusual move, Trejo plans to combine *both* the East

Coast and the West Coast treatment systems—stacking granular acti-vated carbon on top of reverse osmosis, as well as using ultrafiltration, and ultraviolet disinfection, with advanced oxidation. The ultrafiltra-tion will screen out particulates, bacteria and nutrients; the reverse osmosis will screen out pathogens, viruses, pharmaceuticals, pesticides, and "forever chemicals" like PFAS,[18] and the ultraviolet disinfection will tackle any pathogens, viruses, and chemicals that may have slipped through the other treatment layers. At this point the water is so pure that it resembles distilled water. But in an extraordinary additional step—especially on top of reverse osmosis—Trejo is adding granular activated carbon treatment too. He admits that the vast majority of engineers will see that step as unnecessary overkill. What's the point of repurifying purified water? But to Trejo it's all about peace of mind in case something extraordinarily unexpected were to happen at what will be the largest direct potable reuse plant in the United States. "What if? …What if?" he asked. He admitted that he's added these extra layers as much for public perception as for safety. "We wanted to be able to say we're doing everything possible to ensure the quality of the water," he told me. In yet another key step, the employees operating the facility will not only be licensed water treatment plant operators, they will be trained as wastewater operators as well. That step also goes beyond what state officials require.

After hearing Trejo talk about the myriad layers of treatment and redundancies, I asked if this is going to be the most sophisticated water treatment plant in the United States?

"I've never thought about it that way," Trejo replied, "but with the advanced water quality monitoring … and putting these treatment pro-cesses back-to-back-to-back, I think you might be right."

Then, after a pause, he said, "And it should be, considering what we're proposing to do."

That brings us back to the historic drought that hovered over Texas from 2010 to 2015. As bad as that drought was, a scorching dry spell in the 1950s lasted two years longer and for half a century was known as the state's drought of record.[19] The more recent drought caused $10 billion in damages and is known for producing the state's single driest twelve-month period ever recorded (2011). But the earlier drought, which stretched from 1950 to 1957, caused $36 billion in damages and decimated the agriculture sector.[20] According to one estimate, the number of Texas farms and ranches declined from 345,000 to 247,000 during that ravaging eight-year period.[21]

Urban areas suffered too, perhaps none more than Dallas. The city built an emergency pipeline to siphon water from the salty Red River, which makes up much of the border between Texas and Oklahoma. It was enough to help boost Dallas's water coffers at a desperate time.[22] But there were only six hundred thousand people in Dallas County in 1950. More than two and a half million people live there today. Dallas will likely need more than a salty Red River diversion to make it through the next monster drought, which experts say is only a matter of time.[23]

Is Dallas ready for such a calamity? Could a potable reuse program help the Dallas–Fort Worth Metroplex prepare? Daniel Nix thinks so. After what he went through in Wichita Falls, he's one of the state's leading experts on potable reuse. But he admits that there's one complication for Dallas: Houston is highly dependent on treated sewage discharged by the Dallas–Fort Worth metropolitan area. During much of the year, the flow of the lower Trinity River is made up predominantly of effluent from the Metroplex.[24] The river is a key source of supply to Lake Livingston, Houston's primary drinking water reservoir. It is a classic example of de facto indirect potable reuse. People in Houston have been drinking de facto recycled water for half a century. Water-wise Houstonians like to joke that they always flush twice when they are in Dallas to help keep Lake Livingston full. But most of us are downstream from somewhere, including Dallas, whose residents have been drinking at least some de facto recycled effluent for decades too. "Some of the water that we're utilizing has been used by someone already," admitted Denis Qualls, superintendent of planning at Dallas Water Utilities. "It just has a different name now." That means that by the time Dallas's effluent hits Houston, it has likely been de facto recycled more than once. But most people in Houston don't realize that either.

Dallas's water use is projected to rise by more than 50 percent by 2070.[25] Where's the city going to get all that extra water? One additional reservoir will come online by 2030, and although Qualls told me that Dallas has achieved some gains in water conservation, there is still room for improvement, which will further increase the available supply. The city has some limited nonpotable water recycling projects and has considered expanding that program. Qualls said that they may end up dabbling in potable reuse in coming years as well. Potable or otherwise, the Texas Water Development Board sees a lot more water recycling in Dallas's future. Fort Worth's too. "Of the sixteen planning regions within the state, Region C (the Dallas/Fort Worth area) will be the most reliant on water reuse for its water supply," predicted one board report.[26] But if Dallas captures its effluent and reuses it, won't the city be robbing Houston of drinking water?

"We don't see it that way," Qualls told me.

Texas follows western water law, or what's known as the prior appropriation doctrine, which means that people can own water rights—kind of like owning a piece of property. Dallas not only has rights to the water in its reservoirs, but it also owns rights to its effluent. Qualls said that this issue came to a head several years ago when a number of municipalities and utilities in the Trinity River watershed suddenly realized the value of their wastewater and they all rushed to claim effluent water rights at roughly the same time. That sewage scramble led to a "gentleman's agreement," he said, in which everyone in the watershed committed to releasing 30 percent of their effluent downstream, reserving rights to reuse 70 percent—but they could only reuse it once. "So we're not taking all of what Houston would get," Qualls said.

But what happens during the next big drought? When it hits—and he's sure it will—Qualls said that Dallas may need to consider adding potable water reuse to its water supply portfolio. "Will Dallas ever do direct potable reuse? It's possible," he said. "It'll be something that's looked at more in the future.... I want to have as many tools in my toolbox as I can. I never know which one I'm going to need."

After talking with Qualls, I reached out to Yvonne Forrest, director at Houston Water. We talked about the next drought of record and how Dallas might be forced to add potable reuse to its water portfolio, potentially reducing water supplies for Houston. "I think it's a very real scenario," she said, which explains why she works so closely with all the utilities in North Texas to ensure that there are adequate flows in the Trinity River, "even if they started reusing their water." She said that if another 1950s drought were to hit Houston, and the city ended up with just twelve months of water left, Houston's water conservation plan, which she described as "pretty drastic," would be triggered. "I think we could make it through a 2011 [drought], but I'm not sure we'll make it through the full drought of record," she told me. Under such a scenario, she said that Houston would be forced to view its wastewater in a new light. "That effluent water would become much more valuable," she said. "We would have to do more to make sure it was treated and used in our system." Forrest's customer base is 2.2 million and growing. It may be fifty years down the road, but sooner or later she too thinks that Houston, which averages more than fifty inches of rain per year, will need to consider potable water recycling. "We're having these conversations now," she told me. "I know we're going to get there one day. I don't think we have a choice the way the population is expanding."

That is precisely the scenario that Nix thinks Texas officials should be studying. Yes, if Dallas recycled more of its effluent, it could pose a threat to Houston's water supply—unless Houston decided to offset that loss from Dallas by recycling its wastewater as well. "What if they start reusing their water in Houston?" Nix asked. "Do you really need the discharge from Dallas–Fort Worth? That's the question that needs to be studied."

It's an interesting suggestion—to have two of the largest cities in Texas simultaneously adopt potable reuse as a new drought-proof supply. The Texas Water Development Board puts a priority on fifty-year water planning[27] and carves up the state into sixteen different water regions.[28] Nix said that just about every region has a "reuse sliver" in its plan. Since there are a limited number of spots left in the state for new reservoirs, he said that cities will increasingly turn to what some have considered to be "less than desirable" water sources. That includes purification of brackish groundwater, purification of brackish surface water, ocean desalination—and wastewater recycling. Nix told me that many communities will discover that purified sewage can be a surprisingly cost-competitive water supply option. "They're going to look at it more and more," he predicted. "It *is* the future."

CHAPTER 8

Hot Tempers in Tampa

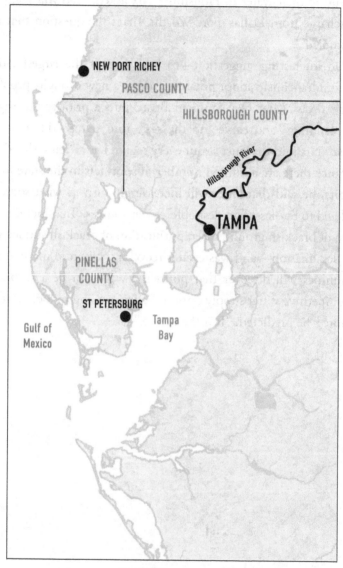

Tampa Bay metro area

IN 1974 MIKE WALLACE, the legendary muckraking *60 Minutes* correspondent, produced an exposé on the unsustainable "explosive growth"
occurring in the Tampa Bay metro area. He reported that an "onslaught"
of seventeen hundred "snowbirds" was moving to the region monthly,
taxing natural resources, especially the water supply. "St. Petersburg ran
out of drinking water fifty years ago," Wallace reported. "They pumped
their wells until they hit saltwater so the only thing to do was pump
drinking water out of neighboring counties, like Hillsborough." In his
classic attack-dog baritone, Wallace added that "there is not enough
water in Hillsborough County anymore" either. He interviewed a
retired veteran beside a disappearing lake who was bemoaning how
his home was now located on the shore of a "mudhole" because of
"that excess pumping." Thanks to the receding waters, the veteran complained, the fish were gone, the birds were gone, and "even the damn
alligators are leaving."[1]

When *60 Minutes* produced that stinging report, the Tampa Bay
metro area was home to one million people. Three times that many
live there now,[2] and the region's water supply seems thrice as stressed.
Yet the transplants keep coming—and not just to Tampa. Today, more
than one thousand people are moving to Florida each day,[3] stressing the
entire state's water supply. Projections show that Florida needs to find
an additional billion gallons of water *per day* by 2040.[4] The Natural
Resources Defense Council said that parts of Florida have a "high risk"
of water shortages by 2050.[5]

Water recycling could help close that gap. Florida has long led the
United States in laying down purple pipe, the color signifying that the
pipe is transporting nonpotable recycled water to lawns, golf courses,
orchards, and industrial facilities.[6] Florida has more purple pipe than Arizona, California, Nevada, or Texas.[7] Expansive nonpotable irrigation systems are common from the panhandle to Key West. It is customary—if
not required—for new Florida subdivisions to have purple pipe irrigation

systems laid from the start, which dramatically lowers the installation cost and discourages people from using drinking water on their lawns.

But with more than four feet of rain per year, why does Florida need to recycle water at all? One reason is that the Sunshine State drains quickly, so the water speeds off into the ocean.[8] That is crucial for Florida's biologically productive estuaries, but it does not help the state with water storage. Generally, only a few inches of rain end up being captured for use later on.[9] Hence, Florida struggles to store large volumes of backup water that can help it ride out severe droughts, like those that struck in 1980, 1984, 1998, and 2006–2008.[10] Florida gets 90 percent of its drinking water from a unique but overtaxed aquifer system,[11] which suffers from saltwater intrusion in many coastal areas. Despite being a national leader in purple pipe, Florida has moved more cautiously toward potable water recycling. Several communities are working on pilot projects that could end up inching their citizens closer to adding purified sewage to their drinking water. But when it comes to potable water reuse, Florida remains far behind states like California, Virginia, and Texas.

No Florida city has tried harder than Tampa to implement a potable water recycling program. In the late 1990s, when San Diego's potable reuse initiative met stiff resistance, Tampa proposed a similar program, which met a similar fate. Tampa tried again in the 2010s—to no avail. In 2021, for the third time in nearly twenty-five years, Tampa water officials launched yet another potable reuse program only to be shot down again—this time by their own city council. It felt like San Diego in 1999. But San Diego has long since moved on from those dark days, investing billions of dollars to ensure that almost 50 percent of its drinking water will come from purified sewage by 2035. In contrast, Tampa is still stuck in the 1990s, fighting the same old fight.

Think of it as the water reuse version of *Groundhog Day*, the irritating Bill Murray film in which the same day keeps repeating itself again and again. The recurrent failures have positioned Tampa in a unique place within the national water reuse community: No city in the United States has worked harder, longer, and more unsuccessfully to implement a potable water recycling program than Tampa. After decades of potable water reuse success stories in places like El Paso, Hampton Roads, Occoquan, Orange County, San Diego, and Wichita Falls, the City of Tampa reigns as the leading persistent example of a community that just can't get its reuse act together. Officials have repeatedly and resoundingly failed to convince local citizens that potable water recycling is a necessary, safe, and effective way to deal with looming water supply woes. A few Florida communities, like Jacksonville and Plant City, have made some notable progress toward potable water recycling programs. But Tampa's situation shows that after decades of incremental acceptance in many parts of the United States, potable reuse projects can still fail—miserably—in other areas. As usual, all it takes is an underwhelming communication plan, a few articulate opponents, and a drought that's just not severe enough.

⌣

Brad Baird has lived through each frustrating episode. He joined Tampa's city government as a young engineer in 1983, eventually working his way up to deputy director of the sewer department and then director of the water department. He oversees both today. Not long after joining city government, his superiors began experimenting with potable reuse. They were heavily influenced by the successes of Water Factory 21 in Orange County, as well as the first potable reuse program on the East Coast in northern Virginia. In the 1980s, the Tampa team even traveled to suburban Washington, DC, to see things for themselves. "We drank the water out of Occoquan," Baird told me.

Nevertheless, Tampa officials still felt compelled to confirm the safety of the emerging potable reuse technology with their own research.[12] They created a pilot facility to test four different potable reuse treatment methods (reverse osmosis, granular activated carbon, ultrafiltration, and chlorine with denitrification). Supported by generous state funding, the pilot program ran for more than two years and included toxicological testing, chemical testing, and biological testing—even some of the nation's first endocrine disruptor research. "You name it, we did it," Baird told me. He said that he and his colleagues felt like they were at the vanguard of the water recycling movement. "It was bleeding-edge testing for the whole world," Baird said. "Except for Occoquan, and Water Factory 21, there wasn't much [potable reuse] going on."

Tampa's research confirmed what Orange County and Occoquan already knew: purified sewage was indeed safe to drink. Of the four different treatment methods, the preferred option turned out to be granular activated carbon—the so-called East Coast method—just like Occoquan. The analytical work was expensive, however, and so was the technology. Although water tensions were high, the Tampa team decided that their local water problems weren't severe enough to justify the cost of investing in a full-blown potable reuse project. The research initiative was shelved in the late 1980s without ever being presented to the public as a formal water supply proposal. "We weren't thirsty enough," Baird said.

But the snowbirds kept coming, with many deciding to stay year-round. The groundwater pumping increased, and water tension did too. There are thousands of lakes and wetlands in the Southwest Florida Water Management District,[13] which includes the Tampa Bay area. Most lakes in Florida are connected to the groundwater system.[14] As groundwater levels plummeted, the lakes did too, infuriating lakefront property owners like the veteran Wallace interviewed for *60 Minutes*. This situation led to what Floridians still refer to as the "water wars" whose foundation was laid in the 1970s and peaked in the 1990s, becoming so heated that a book was even written about them.[15] Water levels in wetlands dropped too, decimating those environmentally sensitive areas.

The water wars pitted counties against one another. As one county withdrew water from underneath another, residents from the water-robbed county became enraged. As one official put it, the overriding sentiment was, "You're taking my water, and you're taking so much of my water that you are damaging my environment."[16] The region's water wars were reminiscent of the parched Southwest. But Tampa Bay is about as far from a desert as one can get: yes, it has a dry season in winter, but it's still drenched with fifty inches of rain per year.[17] Instead, the tensions were driven by rapid growth and groundwater overpumping.

Lawsuits flew. After tens of millions of dollars were exhausted on litigation, a fed-up state legislature pressured the Tampa Bay metro area to create a new regional water authority to stop the fighting.[18] Called Tampa Bay Water, this water supply wholesaler was designed to serve three counties in the Tampa Bay area, as well as three cities within those counties, including the City of Tampa.[19] The idea was to have these different government agencies band together under the umbrella of one regional water wholesaler that would create alternative water supplies and bring relief to the groundwater system. That would, it was hoped, bring the fighting to an end and stop the groundwater declines. Counties and cities sold their groundwater wells to Tampa Bay Water. These government entities would lose direct control of their wells, but as members of the wholesaler, they would have a vote over how that water would be distributed to the region.

There was one caveat. The City of Tampa's main water source was the local Hillsborough River, not groundwater. Since groundwater declines were the problem, Tampa was allowed to keep using the river for the vast majority of its water supply. That meant that the city was generally only dependent on the water wholesaler during extended dry periods, making Tampa a more fickle participant in the new water management system.[20] That dynamic set the table for at least some water tension to continue—often pitting the City of Tampa against some other members of Tampa Bay Water.

As officials settled into this new water management paradigm, the City of Tampa's potable reuse blueprint was pulled back off the shelf. Why? Tampa Bay Water was given $185 million in funding to create additional water supplies. The wholesaler zeroed in on three potential new options: (1) a new ocean desalination plant, (2) a new reservoir, or (3) the City of Tampa's potable reuse idea. The water wholesaler needed to pick two of those three options.

The City of Tampa ramped up its reuse blueprint into a full-blown proposal for the wholesaler to consider. It was called the Tampa Water Resource Recovery Project, or TWRRP. The year was 1997[21]—the same year that San Diego's emergent potable reuse program started to garner headlines. During the next year, the trajectory of the San Diego and Tampa potable reuse projects would become eerily similar. "We were in parallel with San Diego," Baird told me. "We were in lockstep."

Like San Diego, the City of Tampa proposed discharging its purified sewage into a surface water system before it was withdrawn later and treated again as drinking water. There were only slight differences between the two programs. San Diego was planning to discharge into a reservoir; Tampa planned to discharge into a canal. Both programs were classic examples of indirect potable reuse.

Over time, however, public sentiment about TWRRP began to shift. "People started calling it 'twerp,' unfortunately," Baird told me. There were plenty of toilet-to-tap moments as well.[22] But like San Diego, the real killer came in the spring of 1998 when the National Research Council released its skeptical report on potable water recycling, declaring indirect potable reuse projects, like Tampa's, to be an "option of last resort."[23] Fourteen years later—after cataloguing more than a decade of potable reuse success in Orange County and elsewhere—the National Research Council would completely reverse itself by fully endorsing potable water recycling.[24] But that was no help in 1998. If *you* were a regional water wholesaler that needed to pick two out of three new water supply options, and a federal blue-ribbon panel had just declared one of the three to be an "option of last resort," what would *you* do? "We were left at the altar," Baird told me.

So as San Diego's toilet-to-tap controversy went national, the same situation was unfolding in Tampa. But the demise of Baird's project was overshadowed by what was happening on the West Coast. It would mark a striking low point in the national water reuse movement's history: two major water recycling programs on either side of the United States were simultaneously falling apart, hampering the movement for decades. It was the first time that the City of Tampa's potable reuse project was spurned by regional officials, but it would not be the last.

The city began investing heavily in purple pipe instead—a practice that was already widespread throughout the region, especially in St. Petersburg.[25] Baird started adding purple pipe in the neighborhoods with the most extravagant irrigators and worked his way down the list from there. The idea was to convince people to use nonpotable recycled sewage, rather than drinking water, on their grass because the drinking water withdrawals were continuing to put pressure on the beleaguered groundwater system.

It worked—not just in the City of Tampa, but throughout the region. But as Tampa's purple pipe program expanded toward neighborhoods that didn't irrigate as much, the cost of installing the plumbing infrastructure became prohibitive. The potential water savings in these other neighborhoods didn't come close to matching the cost of installation. Things became awkward when a new mayor was elected in Tampa, in part by campaigning on expanding the purple pipe program. Baird's team had to break the news that the mayor's campaign proposal was unrealistically expensive. All the easy purple pipe installations had already been done. One case study showed that it would cost a million dollars to provide purple pipe to just fifty homes, or "twenty grand a pop," Baird said. "From a business case, it was unaffordable."

No worries, the mayor replied, and he asked Baird to come up with some other sexy water-saving idea. "That's how the Tampa Augmentation Project was born," Baird said. It was the City of Tampa's second attempt at creating a potable reuse program. The year was 2013.[26] TAP, as the project came to be known, was a spinoff from the failed TWRRP initiative of the late 1990s. Baird vetted the program with state officials, who "loved the project," he said. He then presented the $260 million proposal to the mayor, who said, "Let's rock-and-roll."

This time Tampa decided to try a groundwater potable recharge project instead of surface water augmentation.[27] But when the TAP program went public, the messaging became muddled.[28] Baird and his new understudy, Chuck Weber, failed to invest heavily in outreach with the community leading to concerns about transparency.[29] They were just two engineers hoping for the best. "We never had the opportunity to do the outreach component for this project," Weber told me. "There are still unanswered, easy questions that are in the TAP reports that if people were to read through them, they'd have their answers—but its thousands of pages. We haven't boiled our messaging down to something that's digestible by the layman."

History has shown that it is virtually impossible to launch a potable reuse project without a raging communications and outreach program. The City of Tampa's TAP project proved to be no exception. The city did hire one of the leading water reuse communications firms in the United States, Katz & Associates out of San Diego. But Sara Katz, the chief executive officer, told me that the relationship was just "transactional." Tampa asked for advice, Katz gave it, and that was that. There was no development of a broader communication relationship or strategy.

There was another problem for Baird and Weber: when they announced TAP, they had not decided how the effluent would be treated.

Would they use reverse osmosis like Orange County? Or granular activated carbon like Occoquan? Or something else? That uncertainty left an opening for opponents to raise concerns about the project, and they did. "On the Brink of a New Water War?" asked the *Tampa Bay Times*.[30]

Then politics got in the way—just like in San Diego. The old mayor was term-limited, and after a local election the majority of the city council was made up of new members, including Bill Carlson, who was particularly critical of the TAP program. Long before then, concerns arose that the City of Tampa was trying to use TAP as a way to gain water supply independence from Tampa Bay Water, the regional wholesaler. The fear was that if the City of Tampa could add recycled water on top of its bountiful river supply, it would never need to call on the regional water wholesaler during droughts. If Tampa gained water independence by adding a new reuse program, regional officials worried the city might then leave the wholesaler. That would fracture the collaborative regional water supply paradigm that was set up to end the water wars, possibly triggering a new era of endless litigation. Seven former officials penned a joint op-ed expressing that fear.[31]

With the old mayor gone, Baird and Weber tried to get Carlson and the other newly elected council members up to speed on TAP, but it was fruitless. The project was seen as a retooled campaign promise from a mayor who was no longer in office. Add in the fears that the City of Tampa was trying to use the added water supply as a way to depart the local water wholesaler, and Baird's second attempt at potable reuse fizzled like the first. In 2019 TAP joined TWRRP as yet another failed potable reuse program in the City of Tampa.

⁓

In 2021 Florida enacted an unprecedented law that is widely referred to as Senate Bill 64.[32] In a remarkable move, the law gave Florida wastewater utilities just eleven short years to figure out how to stop discharging treated effluent into the state's surface waters—unless they could prove that those discharges were somehow "beneficial." To wastewater officials it was an intimidating mandate. Even utility directors who supported the stunning legislation were left rattled by the requirement to reengineer their complicated wastewater systems to prevent millions of gallons of treated effluent from flowing into the state's surface waters. How do you get rid of millions of gallons of effluent—every day? Where else would you put it? As long as anyone could remember, highly treated effluent had been discharged into rivers, lakes, and oceans. It's standard practice. Even the Clean Water Act endorsed the process as long as the effluent met federal standards.

The eleven-year deadline was intimidating to wastewater officials, including Baird, whose city has some of the largest surface water discharges in the state. Utilities generally reacted to the law with two different plans of action. The first was to invest in deep-well injection and send millions of gallons of treated effluent per day far underground—so far down that (it was hoped) it would not have a chance of contaminating Florida's cherished underground drinking water supplies. The other option was to take all of a city's effluent and recycle it—as purple pipe water, potable water, or both.

Senate Bill 64 helped prompt the City of Tampa to revive and rebrand its maligned potable reuse program again. It was the third attempt since 1997. This time they called it PURE.[33] "Out with TAP, in with PURE," wrote the *Tampa Bay Times* in 2021. "It's been more than a year since Tampa gave up trying to sell a plan—dubbed the Tampa Augmentation Project or TAP—to convert wastewater into drinking water in the face of environmental and city council opposition," the newspaper said. "The struggle to persuade critics to get over their aversion to a project they dubbed 'toilet-to-tap' has taken on a new dimension and a new acronym—PURE."[34] The city estimated that the program would cost between $484 million and $628 million.[35]

Yet again, Baird and Weber were wandering through the water reuse wilderness without enough outreach assistance. Within twenty-one months the program was under siege.[36] Opponents claimed that PURE was TAP with new window dressing, even though the program had some key differences. There was an added communication complexity for Baird's team: under PURE, they were leaning toward treating the sewage with suspended ion exchange, a treatment system that had never been used for potable water recycling in the United States. Baird and Weber thought that this method would be cheaper and more sustainable than reverse osmosis. It may have been a great idea, but trailblazing in the absence of a robust outreach effort, with an already leery public, made things exceedingly difficult for the PURE program. "The criticism that we got," explained Weber, "is the only acceptable treatment is [reverse osmosis] membranes. If you're not going to do membranes, then you're not doing the right treatment." The public—and some city council members—revolted against Tampa's experimental tendencies. "Nine out of ten cities that do this use reverse osmosis," said councilman Carlson. "I don't understand why we are experimenting with our public on all this," he told me. "Let's just do what's proven and let some other city be the guinea pig."

As the feeding frenzy built, criticism of PURE became embittered. "This is the third time this project has come back," said the Sierra Club's Gary Gibbons at a blistering news conference in September of 2022. "It's time to kill it."[37] Once again the engineers struggled to find messaging that could compete. "The same thing that happened with TAP happened with PURE," Weber complained. Weber and Baird seemed confounded by the situation they were in—unable to find a way out.

Earlier that year, opponents drew up seventeen key questions and presented them to the city. Months later, Weber and Baird had still not provided answers.[38] "It's a catch-22," Weber said. "We can't answer the questions—until we do the engineering. We can't do the outreach until we have an outreach program that's funded, and we're just not getting that out of [city] council." Baird called it "a brilliant strategy by the opposition." By not allowing funds to go toward a communication strategy, it made it harder for Baird's team to sell the program—and easier for opponents to sow dissension. "[Communication] is one of the most important pieces if not *the* most important piece," Weber said, "and we can't seem to get off the ground with it."

⌣

Nearly two dozen local groups joined together in opposition to PURE. They ranged from homeowner associations to the League of Women Voters and local environmental groups. I was most interested in speaking with the Sierra Club because it had been such an enthusiastic proponent of potable water recycling in San Diego. How could a national environmental organization be in favor of water reuse in one city and opposed to it in another?

I tracked down Nancy Stevens, conservation committee chair of the Sierra Club's Tampa Bay regional office. A retired electrical engineer from Massachusetts, Stevens volunteers for the Sierra Club and in recent years has emerged as a point person on Tampa's water recycling program. "I don't have the background in this, but I've been trying to learn," she confessed. She quickly brought up the now-defunct TAP program. "Since they gave it that lovely name, we called it 'toilet to tap,'" she told me. "The idea was to take the wastewater and put it—not filter it—but to put it actually into the drinking water supply." Weber said that that statement was not true. The city had planned to treat the water, including filtration; Baird and Weber just had not decided what treatment system to use.

Stevens told me, the advocates were trying to make it clear to the city that Tampa was just not ready for potable reuse. The opponents wanted officials to invest more in conservation and purple pipe instead. "We were trying to tell them the public didn't want it," Stevens said. "The public doesn't want to drink toilet water." One of the biggest problems was that the city never conducted the outreach required to convince the advocacy community that a potable reuse program was necessary. "They never have said why they need it. They just think it's a great idea," Stevens told me. "It's not like we have a water shortage here. We have enough water in the City of Tampa."

Really?

That was Stevens's take on TAP. Then she turned to PURE. Starting in late 2020, Baird and Weber met with citizen advocates to talk about the PURE program, but the more the two engineers spoke, the less convinced Stevens became. "I was thinking this is just TAP in a different package," she told me, adding that the new program still had the same unanswered questions. What are the risks? How safe is it? Why do we need it? During the months that followed, the advocates presented Weber and Baird with the seventeen questions.

Things came to a head in September 2022. That's when a request for a million dollars in communications and other funding for the PURE program appeared on the city council agenda. "They had not answered these questions, and so they were going to hire a more expensive marketing firm to make it sound really good," Stevens told me. The advocates did not see the request for communication funding as a genuine attempt to improve the city's outreach efforts but rather as a disingenuous effort to hire a slick public relations firm to sell an unwanted program to the people. Advocates turned out in droves at the next city council meeting to speak in opposition to the program. "At that point," Stevens said, "we were like, 'This has got to stop.'" The strategy worked. The council voted overwhelmingly against the PURE funding.[39]

I was still perplexed as to how a Sierra Club chapter in one state could be so fervently supportive of potable water recycling while a chapter in another state could be so adamantly opposed, including suggestions that the technology was not safe. But to Stevens it was quite simple. "The first thing I did was look at San Diego," she said. Given the Southwest's dire drought, she became convinced that "they need it." But she added that the situation is completely different in the City of Tampa "because we don't need it." She admitted that she couldn't comment on the Sierra Club's national position on potable water recycling because she didn't know what it was, so I reached out to the Sierra Club's national office to find out. But no one there returned my emails and calls seeking comment.

⁓

What I found remarkable in reporting and researching Tampa's potable reuse program is that everyone I spoke with was universally frustrated. Baird and Weber were clearly frustrated. The regional water management district was frustrated. The state water reuse community was frustrated. The advocates were beyond frustrated. Some city council members were apoplectic. "I want PURE to go away!" said Lynn Hurtak during a council meeting in 2022. Waving her arms in exasperation, the councilor said, "I mean—No for PURE! No! Let's just get rid of it!" That Tampa tantrum was not from an advocate but from a leader of city government.[40] Later on in that meeting, the council voted to stop funding PURE for the foreseeable future.

A few months after that vote, the Tampa Bay Young Republicans released a highly critical video of mayor Jane Castor, Baird's boss.[41] PURE had become one of her most controversial programs. Castor is a Tampa native who spent decades as a police officer, eventually rising to chief—the first woman to do so in the city's history. She's also a Democrat, and in December 2022 she was running for reelection.[42] That's when the Young Republicans released their caustic video on social media attacking Castor's PURE water recycling program.

"Have you ever flushed your toilet and thought to yourself, 'I'd really like to drink *that* someday?'" a woman asked sarcastically in the clip.

"Now, thanks to Jane Castor, that dream can become a reality," again, dripping with sarcasm. The woman then places two glasses of brown water on a table.

"Jane Castor's toilet-to-tap program has been shut down numerous times and opposed by just about every environmental group in the state. I can't imagine why?" the woman said. "But thankfully Jane is gonna make this project her number one—or, should I say, *number two*— priority for you, me, and our families," she continued as a fecal emoji flashed across the screen. Then her daughter asked, "Mommy, can I drink my potty water yet?"[43]

The video was a scorching reminder of how vulnerable potable water recycling can become during a political campaign, not to mention for any candidate who supports it. There's just so much sophomoric material there for anyone who wants to stoop low enough to use it.

Channeling all that Tampa frustration, I reached out to Donna Petersen, a professor of public health at the University of South Florida. Petersen had an interesting perspective. She's spent her entire career in public health, but several years ago the Florida water reuse community reached out to her to see if she would peer-review their work—from a public health standpoint—and invited her to join the state's potable reuse commission. She learned a lot about reuse and was won over by the technology. That was the good news. The bad news? She was appalled at the reuse community's ability to communicate. "I'll cut right into my bottom line," she quickly told me. "I don't think the people engaged in this work communicate very well.... These are water engineers, geologists, hydrologists—super smart—but they don't do community engagement and marketing. It's not their thing." She didn't just tell *me* that. She said it to *them* too. "You're not communicating this well and you're doing yourselves a disservice."

She said that PURE was a classic example. Like everyone else in Tampa, she watched it disintegrate. "We had this PURE project. There was support for it. There were funds allocated to it," she said, "but it wasn't communicated well." She said that there were two main critics: the Sierra Club and those she referred to as the "toilet-to-tap group." "The whole thing got torpedoed, which was a real shame." She said that the opponents' entire stance was cloaked in the assumption that "we don't need water."

"Are you kidding?! That's part of the problem," she told me. "People just don't understand water. They don't understand where it comes from."

Due to vaccine conspiracy theories, COVID-19 misinformation campaigns, and other trolling, communications training is a basic course in Petersen's public health program. She's all too familiar with how messaging can go awry. She said that one of the first public health classes in her program emphasizes that if a community doesn't buy into a problem, they won't support a fix. "The community has to recognize the problem exists ... and then agree to a solution," she said. "Because without that, you're just blowing in the wind." Tampa officials have repeatedly whiffed on all those key points.

Then I asked her another question. During the time when she was peer reviewing the state's water recycling protocols, as a health expert, did she ever become concerned or insecure about the safety of the process—from a potable standpoint? "No," she said. "Actually, quite the contrary. I was so impressed with the level of thoroughness, and care, and study, and evaluation." She said that the public is worrying too much about water that is perfectly safe to drink and not worrying enough about the water scarcity in their future. "People will get exercised about an impossible threat and ignore the one that's right in their face," she warned, "which is we may run out of water, people, if we don't figure this out."

I asked Petersen—a health expert who has spent years studying potable water reuse—what would she tell Tampa's potable reuse opponents. "There is no *new* water on the planet.... Every drop we drink has been drunk by someone before and gone through the digestive cycle and the hydrologic cycle. Every bit of it," she said. "You are drinking pee from dinosaurs right now.... We have been recycling and reusing water forever. What's different now is we know there's a threat of running out. We have technology that will allow us to shortcut that recycling system and create a sustainable source of potable water, well into the future. Do we need to continually monitor what's in the water and continually refine our instrumentation to make sure we're removing any impurity that could be a threat to human health? Yes. But the bigger threat to human health is running out of enough water.... Florida would do well to capitalize on all the work that's gone into creating sustainable potable reuse systems and investing in them, now, so that we're not doing it in a crisis."

Tampa was on the brink of a crisis in the spring of 2009. The city was struggling through the third year of a particularly severe drought. Baird told me that in May of that year his seven hundred thousand customers were "within a month of running out of water." The city imposed harsh water restrictions and required residents to turn off their irrigation systems—any watering had to be done by hand. Tampa was buying millions of gallons of emergency water per day from Tampa Bay Water, the wholesaler, "[and] we were still running out of water," Baird said. "We faced the real potential of running out of water totally." Then the rains finally came, and the emergency passed, but not the stinging memory. "This is like yesterday to me," Baird said.

Tampa Bay Water disputed that account and said that the city still had water options.[44] So did the head of the Southwest Florida Water

Management District.[45] What everyone can agree on, however, is that another big drought is bound to happen.

I asked Baird if the City of Tampa was ready for it.

"No," he answered.

"And what's going to happen?"

"We face the scenario of possibly running out of water."

"Seven hundred thousand people?"

"Correct."

"And who's to blame?"

"I don't know.... I'm not sure that anyone's to blame," he said. "You have to be solving a problem and right now nobody thinks we have a problem."

"And could a large potable reuse project prevent that crisis from happening?"

"That is correct," he said.

I then asked how worried he was about that crisis scenario.

"I'm worried—very worried," he said. "I mean water is everything. Water is life. Without water, you die."

⌣

By the spring of 2023, the resounding criticism of PURE had taken its toll. Tampa terminated the program and scrubbed it from the city's website. Weber took a job at a different utility, leaving Baird without his understudy as he contemplated the city's next move. The death of PURE left Tampa with a potable reuse record of 0–3.

I walked away from Tampa as frustrated as everyone else. I also walked away thinking that the entire region could stand to make long-term investments in potable reuse. Many of the people I interviewed admitted that they were not opposed to potable water recycling—in principle—some just didn't like how the City of Tampa was going about it. It seemed as if reuse had become collateral damage in the water turf battles that have been a signature of the Tampa Bay region for decades—dating back to that 1974 *60 Minutes* report.

But it also struck me as one of those cases where fighting makes everyone look bad. The water wars may technically have ended when Tampa Bay Water became the regional wholesaler, but my visit showed that there was still plenty of post-détente skirmishing going on, with water recycling taking the biggest hit. Given the bitterness, bad blood, and entrenched positions on all sides, Tampa may remain stuck in the 1990s for the foreseeable future. After all, it took San Diego decades to bounce back after its program was killed by the city council in 1999.

The various parties in Tampa were still talking as this book went to press. I hope they can figure things out before the next big drought.

CHAPTER 9

Going Beyond Purple Pipe in Florida

Florida

DESPITE THE REPEATED FAILURES IN TAMPA, other Florida cities have found the transition to potable reuse to be smoother, at least so far. Take Jacksonville, for example, Florida's largest city. At just under a million people, it is bigger than Miami and Tampa combined. It takes thirty-eight water treatment plants and eleven sewage treatment plants to keep all those people watered. In 2014 the city's utility, JEA, embarked on a long-term water planning exercise with the local water management district. The study did not bring good news: JEA, which gets 100 percent of its drinking water from underground, discovered that it would start bumping up against the limits of its permitted supply by 2030. It needed to find more water.

But it's not just Jacksonville. One legislative study declared the region surrounding the city to be the most water-insecure in Florida—needing to find 140 million gallons of additional water per day by 2040.[1] What's more, 67 percent of the entire state has been designated what is called a Water Resource Caution Area.[2] The state defines these areas as "having existing water resource problems" or areas where "water resource problems are projected to develop during the next twenty years."[3]

Having already invested heavily in conservation and purple pipe, JEA examined thirty different potential water supply options, eventually whittling things down to ocean desalination, potable reuse, or local surface water supplies. Additional research showed that potable reuse would be cheaper, easier, and more environmentally friendly than the other options, especially ocean desalination. Why? Because the waters off Jacksonville are environmentally sensitive. For one, they represent the heart of the calving grounds for the North Atlantic right whale,[4] which has been on the federal Endangered Species List since 1970.[5] With fewer than seventy breeding females left,[6] those waters did not seem like the best place to install a desalination intake pipe, not to mention disposal of brine discharge.

But Jacksonville is also located on the St. Johns River, a major waterway. Why not rely mainly on surface water like Tampa does? On an annualized basis, surface water turned out to be more expensive than potable reuse, in part because tidal influences increased the salt content that would need to be removed from the river water.

Ryan Popko was selected to lead this potable reuse program, which JEA calls H2.0, as in "H two point O." After earning two environmental engineering degrees, the New England native ventured to Florida in 2005 looking for a water supply adventure. He wanted to make a difference. "When I graduated, I decided to move to a state where there were a lot more water challenges," Popko told me, "so it was kind of between Florida, Texas, and California." Florida won out, and he first worked as a consultant in Jacksonville before landing at JEA, where he finally found his water challenge. After the utility decided to pursue potable reuse, Popko did his homework by visiting several successful water recycling programs across the United States, including Hampton Roads, Virginia; Orange County, California; and San Diego. The goal was to tease out lessons learned and get schooled on recommendations about how best to start a successful potable water recycling program.

What was his key takeaway from those various field visits? "Customer engagement and education is essential," Popko told me. "The technology is the easy part. But if you can't get your community on board with what you're doing—and if your community does not have faith in you as a utility—then you're going to struggle in getting a program like this off the ground."

He also learned that several of the most successful potable reuse programs followed a simple three-step process. First, construct a small pilot plant to test the technology, which can help build public trust. This phase also includes outreach to key public figures in government, business, and the nonprofit sector. Assuming all that goes well, step two entails building a one-million-gallon demonstration facility to allow the public to come learn about the technology and even sample the water. This multiyear phase also includes an extensive public outreach and education campaign. "You have to be transparent. You have to tell them exactly what you're doing and why you're doing it and be able to take a really complex subject and make it simple," Popko said. The third step—again if all goes well with steps one and two—is to build a full-scale water purification plant capable of producing millions of gallons of purified sewage daily. The entire process can take well over a decade.

Popko's team completed step one in 2019 and through that process selected reverse osmosis, with groundwater replenishment, as the primary treatment system (like Orange County). Step two is well underway. When the demonstration plant is completed in 2025, it will include a large visitors center. Popko said that he expects to host thousands of people per year—from high schoolers on up, "because we're also looking at the future leaders of our community." The facility will help explain "why we need to do this, teach them about the water cycle—that all water on Earth is recycled now—we're just doing what nature does ... in a smaller footprint and a shorter amount of time," he told me. JEA had started investigating potable reuse back in 2014, so by the time the state decided to ban all nonbeneficial sewage discharges in 2032, Popko's utility was well positioned to meet the demands of that legislation without panicking. He admitted that the 2032 timeline was "aggressive," but his utility supports the legislation. "Fortunately, we started our plan quite a while ago, so we're a lot farther ahead than most other utilities," he said.

After hearing Popko quickly and clearly rattle off the various steps of the JEA potable reuse strategy, I couldn't help thinking back to the City of Tampa. Yes, Tampa did a pilot project—in the 1980s. What's more, the city's last major communication program was for TWRRP—in the 1990s. From a public engagement standpoint, that's ancient history. With PURE, Tampa was skipping over many of the important outreach stages that JEA, El Paso, Hampton Roads, and others have shown to be an integral approach. Perhaps most difficult of all for the public, the City of Tampa was leaning toward using suspended ion exchange to purify its sewage. Ion exchange is a technology that is likely to be safe, cost effective, and sustainable—possibly even more sustainable than reverse osmosis—but it has not been used for sewage recycling in the United States before. It may be best to leave that experiment, no matter how worthy, to a city whose water reuse program is not mired in controversy already. "Sometimes it's easier to install the 'tried-and-true,' and try to optimize from there," one senior Florida water official told me, "as opposed to blazing a new path."

JEA's initial success in building a potable reuse program is not an isolated incident. Several other Florida communities, including Altamonte Springs, Daytona Beach, Plant City, Polk County, and Hillsborough County, have shown similar promise, although their programs are smaller than what JEA hopes to build. But thanks to the troubles in Tampa, many of these communities are proceeding with caution—and perhaps even a sense of anxiety. As one Florida water official told me, "I had the fortunate experience of seeing all that didn't work with Tampa and saying, 'Here's what I'm *not* going to do."

But there's also a feeling of urgency among water utilities in the state. That's partly due to Senate Bill 64's impending ban on non-beneficial sewage discharges, but it also comes from a concern about Florida's pressing water supply needs. "Our state has challenged us to do this very quickly," said Lynn Spivey, past president of WateReuse Florida, who is spearheading a promising potable water recycling program in Plant City, east of Tampa. "Portions of the state are already seeing concerns with future water use," she told me. "It's getting real." Spivey said that the state needs to do more, sooner, in terms of creating additional water supplies. "We are so behind," she said. "I'm glad I feel a sense of urgency instead of the general public, because I think if the general public truly understood, there would be panic."

Spivey moved to Florida as a teenager. Like Popko in Jacksonville, she spent years working as a consultant before crossing over to the municipal side. "To me, water treatment was always about getting water back to its natural state," she said. "I jumped right into recycled water because that seemed to be where the most work was needed." Florida has been the king of purple pipe for a long time.[7] Now the question is whether Florida can go beyond purple pipe by equally embracing potable water recycling as well. "We need other tools in the toolbox," she told me, "Indirect and direct potable projects are the next tools." But she said a key problem is that most Floridians don't think that potable reuse is needed because they don't realize that the state is water-stressed. "California can see the need for water," she told me, but Floridians "don't look around and feel thirsty.... Florida doesn't have that sense of urgency that we need water."

Tampa aside, there is a lot of cautious optimism and positive energy about potable reuse within Florida's utility industry. The legislature has taken several actions in support of reuse.[8] There are some promising signs of public support too—at least outside Tampa.[9] To Bart Weiss, a water reuse expert at Hillsborough County, it shows that the state is starting to see recycled water as something that is too precious to just use on lawns. "Now with potable reuse coming about," he told me, people are saying "'Hey, this resource is valuable.... We should stop throwing it away.'"

Hillsborough County has used recycled water to recharge a local aquifer since 2015. Called the South Hillsborough Aquifer Recharge Project, or SHARP, it not only boosts the groundwater supply, but it pushes back saltwater intrusion as well. It is the first aquifer recharge project in the region, pumping an average of fourteen million gallons of recycled water underground per day.[10] Even though SHARP is just down the road from Tampa, it has managed to avoid the negative publicity that has hounded Tampa's water recycling efforts. There's a reason for that. "We didn't get a lot of controversy," Weiss told me, "because nobody uses the aquifer." It's another example of how Florida is tiptoeing its way toward potable reuse.

Weiss is a seasoned hand in Florida's water world, having worked for a number of agencies in southwest Florida during his career. Throughout that time, he has watched politicians use water controversies for their own political gain. "If you could just do one thing and take the politics out of water," he told me. "But you'll never get it out. It's unfortunate. It's always been there, always will be."

⌣

Ryan Popko, up in Jacksonville, told me that what Florida needs is a resounding potable reuse success story. An accomplishment like that would help convince state residents that potable water recycling is a safe and effective way to navigate future water insecurity and help Floridians move beyond their long-standing reliance on purple pipe. "What the state needs is a leader like Orange County out in California," he told me. "All Florida needs is someone to take the lead and create the path that everybody else can follow."

"Is that JEA?" I asked.

"I'd love to be able to say that," Popko told me. "[Orange County] has provided such a role model of what you can be in this industry. Folks like them, and Hampton Roads, are definitely what we strive to be."

There's no doubt that a successful program in Florida could help pave the way for a potable reuse rally. The central Florida community of Altamonte Springs (population forty-six thousand) was one of the early adopters. The city's pureALTA pilot program uses granular activated carbon (East Coast method) to produce twenty-eight thousand gallons of purified water per day. At the moment, the recycled water is used for irrigation, but the city envisions the pureALTA pilot paving the way for a much larger plant that could augment future drinking water supplies.[11]

A success at JEA would likely garner a lot more attention. Others argue that a smattering of small communities may do just as much to set an example for the state. Spivey, who is building the potable reuse program in Plant City (population forty thousand) thinks that smaller communities may be able to move faster and more nimbly than bigger cities like Tampa or Jacksonville. She said that Tampa needs to worry about seven hundred thousand people. She only needs public outreach for forty thousand. "We're doing something to help our water supply," she said. "We're letting folks know we are setting the bar. This town is really proud of that.... That's the kind of backing you need."

Outside Tampa there is a quiet competition going on among a host of Florida utilities. They are collegially competing to see who can build the state's first full-fledged potable reuse program. "Plant City may be the first one, because they're doing it right," predicted Brian Armstrong, executive director of the Southwest Florida Water Management District. "They're marketing and selling their project now, before they need it." Spivey agreed. "We think we can set the example of how potable reuse can be done—all the way from how to pilot it, how to permit it, and how to work with the Department of Environmental Protection," she said.

"Florida will never run out of water. We're a peninsula. We're surrounded by it," added Armstrong. "We've just run out of cheap water. We can desal 'til the cows come home, but people aren't willing to pay for that." To Armstrong, that makes potable reuse "the next lowest hanging fruit." He told me that some counties and cities are holding off on installing additional purple pipe projects because they want to reserve that water for future potable reuse instead. "They recognize there's this next bigger thing out there that actually could be revenue generating and could help meet water supply demand.... It's really, how do we sell it to the public?"

Large or small, momentum is building for potable reuse in Florida. Potty water jokes aside, the state's water reuse community finally seems poised to move in coming years from the purple pipe era to the potable reuse era. As numerous utility directors continue to advance methodically in that direction, they have one simple request of their fellow Floridians: judge our water by its quality, not by its history.[12]

CHAPTER 10

LA Goes All In

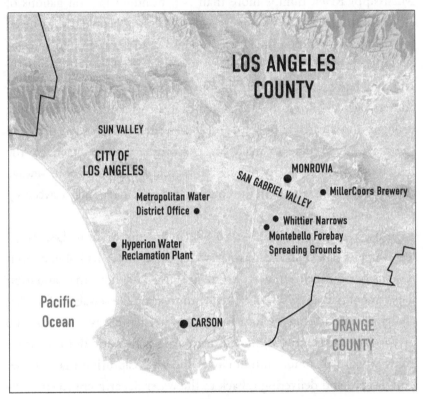

Los Angeles

WITH SWEEPING VIEWS OF SANTA MONICA BAY as a backdrop and waves softly breaking on the beach behind him, Los Angeles mayor Eric Garcetti stepped up to the podium. Given the setting, it was hard to tell that he was speaking from the city's famed Hyperion Water Reclamation Plant. The plant, which is the largest sewage facility west of the Mississippi River, dumps more than two hundred million gallons of effluent into the Pacific Ocean every day. Scores of people had gathered for the mayor's speech, including top water and wastewater officials, as well as leading environmentalists. The date was February 21, 2019, and they had come to witness the most transformative water recycling address ever delivered in the United States. After years of punishing drought and a cascade of state and local water conservation dictums, the mayor had come to make an extraordinary announcement: the nation's second largest city planned to recycle 100 percent of its sewage by 2035. No community in the country had ever made such a sweeping potable reuse pledge.

In his classic laidback style, so befitting a mayor from Los Angeles, Garcetti talked about how the water department had always been responsible for delivering water to the city and how the sanitation department had always been tasked with treating the sewage before discharging it into Santa Monica Bay. The water has always come in, the mayor explained, and the effluent has always gone out. "But now we're going to turn this more into a circle," he said, capturing that sewage, purifying it, and delivering it back to the water department so city residents could drink it. In the process, Angelenos will be recognizing "what nature has already taught us," the mayor said. "That all water has always been recycled."[1]

The program was dubbed "Hyperion 2035." The scope of Garcetti's vision was extraordinary: a sixteen-year, multibillion-dollar initiative to produce more than 200 million gallons of recycled water daily.[2] When complete, it will be the largest potable water recycling effort the United

States has ever seen. In a mind-boggling engineering endeavor, the project will thoroughly transform the way Hyperion's sprawling 144-acre campus processes sewage. Most impressive of all, the wastewater plant will be retrofitted into a reuse facility—*while it continues to operate*—treating hundreds of millions of gallons of effluent daily.

To water pundits, there was another notable twist. Neither Garcetti nor anyone else on his team said anything about purple pipe. In LA, recycled water had become too valuable to waste on people's lawns. They wanted to drink it instead. According to the plan, purified effluent will eventually supplement nearly half of the city's drinking water supply. "It's hard to overstate how important this announcement is for Los Angeles," one city councilmember said. "Big things can happen in big cities when we work together toward a sustainable future."[3]

The announcement had significance far beyond Los Angeles. It was transformative for the national water recycling movement as well. In a historically significant moment, a major political figure confidently strode up to a microphone to tell millions of people that in coming years they will be drinking their own purified effluent. There was no uproar, no whining, no fearmongering, no sky-falling screeds. It was just another major water story in a thirsty city searching for additional supplies.

The *Los Angeles Times* called the program "ambitious."[4] Others called it "bold."[5] But none of the news stories tracked down any "Revolting Grandmas" for sensational comment nor were there any juvenile "potty water" videos. The only person to mention "toilet to tap" was Garcetti himself as he reminisced about how the moniker was invented years ago to demonize water reuse. But those days are over, Garcetti suggested. LA's water situation was so dire that people weren't afraid of water recycling; they were afraid of running out of water. There will always be some who doubt the technology,[6] but thanks to unwavering support from LA's influential environmental community, most Angelenos greeted Garcetti's news with something akin to a collective shrug.

Potable water recycling was becoming normal in Southern California. "Recycling water that is currently discharged to our oceans isn't just achievable," one advocate said. "It's necessary."[7]

But wait, there's more. Garcetti was not the first to think big about potable reuse in Los Angeles. That credit goes to the Metropolitan Water District of Southern California. Serving nineteen million people and twenty-six public water agencies, it is the largest drinking water wholesaler in the United States. Metropolitan has long been known as the king of water diversions, piping in almost two billion gallons per day from northern California and the Colorado River. But the worst drought in twelve hundred years has forced the wholesaler to revisit that model. Climate change has proven water diversions to be an increasingly unreliable source of supply. Starting in 2008, Metropolitan reached out to the Sanitation Districts of Los Angeles *County*, which owns a sewage treatment plant almost as large as Garcetti's Hyperion plant. Together the Sanitation Districts and Metropolitan developed similarly ambitious plans to purify 150 million gallons of sewage per day. They were already a few years ahead of Garcetti's team when the mayor made his historic announcement in 2019.

Both of these water recycling programs are enormous on their own, but when added together they represent a gargantuan water reuse investment for the LA metro area. Combined, the programs are poised to deliver more than 350 million gallons of purified effluent per day. That will turn Los Angeles into the potable water recycling capital of the world. When you add in Orange County's current record of 130 million gallons, not to mention another 83 million gallons in San Diego—and other smaller potable reuse programs in the region—Southern California is becoming the Silicon Valley of potable reuse. In fact, so much water recycling investment is planned for the LA metro area that people are starting to wonder if there will be enough engineering contractors to do all the work—or enough trained operators to run the new high-tech facilities. In short, Los Angeles has become a water recycling boom-town. No metro area in the United States has staked so much of its water future on potable reuse. "If they're able to pull this off, these two big projects will basically replumb Southern California through the lens of water recycling," said Patricia Sinicropi of the national WateReuse Association. "That will probably lead to a tipping point toward more communities adopting the practice.... It's huge."

⏜

What's surprising about LA's water recycling history is how far back it goes. The metro area's reuse story is as venerable as that of any other place in the nation. The problem—for the city and the county—is that not all that history has been positive. When you consider how pessimistic things became in the 1990s and early 2000s—when two major potable water recycling initiatives died in a maelstrom of misinformation—it's almost miraculous that LA has turned into the potable reuse hotspot that it is today.

It all started with a burst of enthusiasm during the Great Depression. In 1930 Los Angeles opened California's first potable groundwater recharge pilot plant. The West Coast plant used what is now known as the East Coast method of water recycling: granular activated carbon. The "polished effluent" was then sent to nearby spreading basins—large dry pond beds—where it percolated into the groundwater.[8] Some five hundred people toured the facility, many sampling the product. Plans were drawn up to expand the demonstration plant into a full-fledged twenty-five-million-gallon-per-day operation. Then it all fizzled. Historians hypothesize that California's seminal water recycling program died after Metropolitan secured water from the Colorado River.[9] At the time, imported water looked more attractive than recycled.

In the decades that followed, nonpotable purple pipe programs spread throughout Southern California, including the Los Angeles area.

History was made in the 1960s when the Los Angeles County Sanitation Districts created the state's first permanent potable reuse program. For years the LA region had been capturing stormwater flows and importing water from the Colorado River and discharging that water into spreading basins where it would soak into the ground, recharging the local aquifer.[10] But in 1962, the Sanitation Districts tweaked that tradition by building a water reclamation plant near the Montebello Forebay Spreading Grounds at Whittier Narrows, east of downtown LA. The program was developed in partnership with the Los Angeles County Flood Control District and the Water Replenishment District, which was specifically formed to manage groundwater in the area.

What was different about the Whittier Narrows program? It sent ten million gallons of treated wastewater per day to the spreading grounds.[11] The initiative was seen as an important additional local water source while LA was waiting for the emergent State Water Project to start diverting new supplies from Northern California.[12] The state project, which came online in phases in the 1960s and 1970s, was designed, in part, to augment supplies piped in from the Colorado River.[13] Despite the groundbreaking nature of the Whittier Narrows water recycling project, there was no fuss. "We could never have known if reclaimed water was acceptable," said one official in 1964, "unless the plant had been built."[14]

That was a decade before Orange County would launch Water Factory 21. Unlike Water Factory 21, it's important to note that the Whittier Narrows plant did not use reverse osmosis to treat the water before sending it to the spreading basins. Rather, it merely treated the water to what is known as secondary standards. (In the 1970s that treatment system was upgraded to tertiary standards[15] with filtration: clean enough to swim in, but not approved for drinking.[16]) Instead, the Whittier Narrows program left it up to the percolation process, and residence time in the aquifer, to transform the treated effluent into drinking water.

Compared to Water Factory 21's reverse osmosis approach, the Whittier Narrows method could be considered something akin to "water recycling light." Yet it worked. During decades of regular monitoring, officials insist the groundwater at Whittier Narrows repeatedly received a clean bill of health.[17] That raised an interesting question for the water recycling community: is Orange County's reverse osmosis system overkill? Or after more than a half century of percolating, is Whittier Narrows still just lucking out?

That question would become front-page news in Los Angeles during the 1990s, a decade that started amid the worst drought since 1977's "near-Saharan summer."[18] As the 1990s dry spell dragged on, Earle Hartling emerged as a key water recycling figure in the Los Angeles metro area. Hartling joined the Los Angeles County Sanitation Districts straight out of grad school, and eventually became their water recycling coordinator. As the 1990s drought was ramping up, he reached out to the Upper San Gabriel Valley Municipal Water District in Monrovia, about twenty miles northeast of downtown Los Angeles. He was pitching an expansion of the Sanitation Districts' potable water recycling program and proposed construction of a nine-mile pipeline to help recharge San Gabriel Valley's aquifer. Just like Whittier Narrows, the idea was to discharge reclaimed water into local spreading grounds so that it could trickle into the soil. San Gabriel Valley officials liked the idea, and Hartling went to work.

Miller Brewing didn't like it all. As is the case with most brewers, Miller goes to great lengths to ensure that its beer is made from the highest quality water. Its regional bottling plant brewed more than five million gallons of beer per year—with water pulled from the same aquifer that Hartling was targeting. No matter how safe Hartling said the water was, the brewer was appalled with his plans. "Miller—if I may use this expression—lost their shit over this," Hartling told me. "Their lawyers came in and threw everything but the kitchen sink at this project."

Then a second front emerged. Hartling found himself battling with Forest Tennant as well. Tennant, a physician, ran a series of medical clinics in LA, and he printed ads in regional papers claiming, without evidence, that Hartling's project would cause everything from birth defects to dementia.[19] Tennant was also a former mayor in West Covina, which made him a well-known political figure in the San Gabriel Valley. The doctor said that recycled water should only be used for irrigation or industrial applications.[20] "They've got to include reverse osmosis," he said, "before I'll support it."[21]

Then a third front emerged. This time Hartling was squaring off with a clown. Literally. "Full clown outfit, the grease paint," Hartling told me, "the whole Bozo stuff, big feet." He first met E.T. the Clown at a public hearing about the water recycling project. Things started out nicely enough. "He said, 'My name is E.T. the Clown and I do this cable access show and I'd really like to talk to you.'" To his eventual regret, Hartling spoke to the clown, briefly, before the start of the hearing. During the meeting's public comment section, however, the clown's tone changed. "He went absolutely nuts on this project," Hartling told me. "He claimed that this was—and I'm going to quote him—'a conspiracy by the Trilateral Commission to poison the people of San Gabriel Valley.'"

Hartling brought a Whittier Narrows water sample to the hearing to show people how clear it was.

E.T. the Clown pointed at Hartling and said, "If this water is so good, I'd like to see this guy drink it!"

"So, I did," Hartling told me. "Right in front of him.... That's when something snapped in this guy, and I became his personal Satan." Remember that Hartling's water was only treated to tertiary standards—generally considered safe enough to swim in, but not safe enough to drink. The idea was to have residence time in the aquifer purify it further. But over time Hartling became somewhat famous for drinking it anyway. E.T. the Clown shadowed Hartling at other hearings. He even called Hartling's boss to file a performance complaint, which Hartling's supervisor ignored.

E.T. the Clown, whose real name is E. T. Snell, remembers those days fondly. In an interview for this book, he gladly confirmed Hartling's recollections. I asked him why he became a clown activist and what concerned him about the recycled water program. He described himself as an "ex con—no sex crimes, no drugs," six foot four, and tattooed. As a convicted felon, he was not eligible to run for office, but he still wanted to make a difference. He worried that recycled water would harm children, so he did "whatever I could do to harass [Hartling]."

The multifront controversy over Hartling's project would end up leaving a permanent scar on the water recycling industry. The negative publicity generated what is believed to be the first documented use of the term *toilet to tap*. Yes, in the late 1990s San Diego launched the moniker into the national lexicon, but the expression was born in a Los Angeles County newspaper advertisement during the fall of 1993. "That was the first time it was used in the press," Hartling told me, "in the *San Gabriel Valley Tribune*…. And it was accompanied by a picture of a toilet, a picture of a pipe leading to a tap, emptying into a glass of water." The ad was signed by Dr. Forest Tennant.[22]

But Miller proved to be Hartling's most formidable adversary, eventually filing suit against the project. "The imagery of pristine lakes and streams has helped sell American beer for decades," said a *Los Angeles Times* story about the litigation. "So what is a brewer to do when it finds out some of the water it uses to make beer is going to come from the sewer?"[23] The suit claimed that Hartling's project would "irreversibly pollute" the groundwater basin. "The only question is how disastrous the contamination will be."[24] Local officials countered that the water was safe and that Miller's primary concern was bad publicity. "Made from sewer water," one official said, doesn't really resonate with the pub-crawling crowd.[25]

Miller denied the charge and published a full-page ad in the *Los Angeles Times*. "Setting the Record Straight," the ad read. "Miller Brewing Company wants you to know that there are still serious and significant questions remaining about the Upper San Gabriel Valley Sewage Recharge Project," which was not the program's official name. "It is unfair and inaccurate to portray Miller as obstructionistic or merely self-interested in regard to this project." The ad was clearly designed to position Miller as a defender of water quality for the public. "Miller Brewing Company is proud to take a stand for water quality and your right to know."[26]

If Miller was trying to avoid bad publicity, it didn't work. Jay Leno started mocking the controversy on *The Tonight Show*. He asked if beechwood aging would soon be replaced with "porcelain aging" in the beer industry.[27] Despite that negative publicity, Miller's lawyers scored a major win: convincing a local groundwater board to back away from the water recycling project. "This isn't only a victory for Miller," crowed the company's attorney. "It's a victory for the health of the valley."[28] Then Miller scored another legal victory, successfully convincing a judge to void the project's Environmental Impact Report (EIR). After spending $400,000 fighting Miller, in the wake of those two losses, the Upper San Gabriel Water District backed down. "It was really the voiding of the whole EIR—that's what did it," Hartling said. "EIR's are expensive, and extensive, so having to redo it was just too much." Hartling's operation was scaled back to a demonstration project, a move that both Miller and Tennant accepted.[29] "[The District had] just lost too much money, too much political will," Hartling told me. "And they said, 'Okay, we're not going to do this.'"

As Hartling was still recovering from his water war wounds, Orange County reached out. Yes, Water Factory 21 had been a resounding success, but officials there were in the process of increasing their groundwater recharge program dramatically with the launch of the new

Groundwater Replenishment System. They asked Hartling and a colleague to brief them on what went wrong. After walking the Orange County officials through the various highs and lows, Hartling decided to leave them with one additional thought. "'Everybody put your hand up in the air,'" he remembers saying. "'Reach behind you and locate your spine, because you're really going to need that.'"

⌣

While Hartling was fighting his battles in Los Angeles County, the City of Los Angeles was developing plans for its own new potable reuse plant. LA's water recycling initiative started around the same time as Hartling's. The idea was the same too: release treated effluent into a local spreading basin to recharge the aquifer as had been done in Whittier Narrows for decades. The water recycling project would pipe water to the city's Sun Valley neighborhood. They called it the East Valley Water Reclamation Project. The city held uneventful public hearings about the project, which were covered by the *Los Angeles Times*.[30] Some residents raised concerns, but there was nothing close to the organized opposition—at least at first—that Hartling experienced in Los Angeles County.[31]

Environmentalists loved the city's East Valley project, which helped bask the effort in positive media coverage. They were particularly thrilled that the initiative would offset, at least in part, a controversial water diversion from the Mono Lake watershed that had supplemented LA's drinking water for decades. In the absence of any organized opposition and with the Sierra Club and the Mono Lake crowd at their back, the city council held a press event in 1995 accepting a check for millions in federal funding for the new project.[32] During the next five years, city water officials spent $55 million quietly building the East Valley project. During that same period Hartling's program disintegrated.

Then, in the spring of 2000, amid a hotly contested mayoral race, LA water officials announced that the East Valley project was about to go online. The move surprised residents who had not heard about it for half a decade. "This is human waste," one homeowner complained. "I'm very uneasy about that."[33] As more and more LA residents realized that their homes were in the new water recycling service area, opposition grew, steamrolling proponents from the Sierra Club and Mono Lake contingents.[34]

Mayoral candidates tried to outdo one another in criticizing the East Valley project. The *Los Angeles Daily News* covered the story intensely; the *Times* less so. A leading critic was Joel Wachs, a city councilman who had approved the water recycling project years before, but now opposed it as a mayoral candidate. (He said that he didn't realize what he was voting for the first time around.)[35] "It's enough to make you gag," he wrote in one op-ed. Even though the $55 million project had already been built, Wachs insisted that it be suspended until local, state, and federal officials could testify at an investigative hearing.[36] Other mayoral candidates opposed the project as well, including James Hahn, the city attorney, and Antonio Villaraigosa, an LA assemblyman. As the *Los Angeles Times* would later put it, the city water department "could not have chosen a more inopportune moment" to launch the project.[37] Miller Brewing even asserted itself, briefly, into the East Valley controversy, but with little effect. Its brewery was miles away from the East Valley aquifer.

It was too much for Jay Leno to ignore. That's not surprising given that at the time *The Tonight Show* was based in Burbank, just outside Los Angeles. For Leno, this was a hot local story, but he made sure that it went national.

"How disgusting is that?" he said of the reuse program. "You thought it was bad when just the air was brown in L.A."

"You know the worst thing about drinking this water?" he continued. "As you swallow it, as it goes down your throat, it automatically does that swirling thing."

"I saw some environmentalists … on TV last night, and they said recycling toilet water makes a lot of sense," Leno quipped. "They said it would save water—which is true. Of course it will save water—no one will drink it."[38]

Reuse proponents cringed—even those who lived far beyond Los Angeles. Sara Katz was San Diego's lead water reuse communication consultant. She had watched her city's program shatter in a toilet-to-tap feeding frenzy the year before. As bruising as that experience was, she suddenly felt grateful that her program managed to avoid Leno's biting monologue. "You had Jay Leno up in LA saying, 'Oh my gosh, California's lost their mind. They now want us to drink our own toilet water'—and then he would have the sound of wooosh!" she remembered, referring to the sound of toilets flushing.

James Hahn, LA's city attorney, became mayor in 2001, and he promptly killed the East Valley potable program.[39] Following that move, water from the $55 million project was slated for irrigation and industrial purposes—not for the potable water recycling for which it was built.[40] Hahn's decision marked the final blow for Southern California's once-promising potable reuse revival. Between Hartling's failure in Los Angeles County, San Diego's water reuse implosion a few years later, and the East Valley debacle in 2000, three major Southern California potable reuse programs tanked in five years. It was a depressing time for the water reuse community.

During the next several years, potable water recycling went dormant in the metro area, but the drought kept getting worse—with 2007 clocking in as LA's driest on record.[41] (That record would be repeatedly broken in subsequent years.[42]) What's more, an Endangered Species Act court decision substantially reduced deliveries from the State Water Project.[43] The drought was getting worse as water supplies were being curtailed by the courts. Water stress was rising throughout the LA metro area.

By that time, Antonio Villaraigosa had become LA's mayor. In an awkward historical moment, he proposed a potable reuse project of his own, just eight years after opposing East Valley.[44] "If we don't commit ourselves to conserving and recycling water," he said, "we will tap ourselves out."[45] As one article put it, "The one-time critic has become a leading proponent for purifying sewage."[46] He wasn't the only high-profile opponent to experience a change of heart. Forest Tennant, who helped Miller Brewing kill Hartling's project, flipped too. "We can now confidently say to the public that this is safe," he said in 2008. "We know we have a shortage of water coming, so it's time to bring back reclaimed water and I fully support it."[47] He wasn't calling it toilet to tap anymore. Villaraigosa's reuse conversion led to more pilot analysis and testing, but no major potable reuse projects were built on his watch.[48] There was a sense of urgency, but not much action. The metro area, it seemed, was waiting for a new generation of water recycling leaders who had the courage to think bigger.

CHAPTER 11

Pure Water SoCal and Operation NEXT

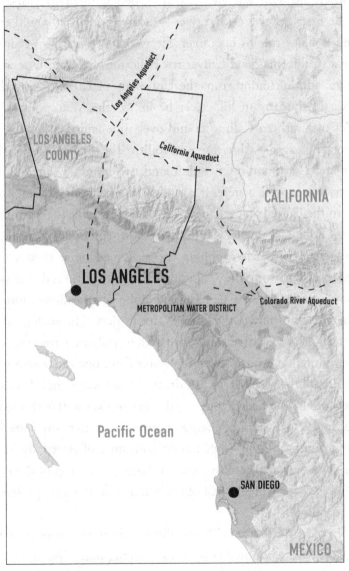

Los Angeles County

DEVEN UPADHYAY FIRST BECAME ACQUAINTED with the Colorado River as a young man. Raised in California and Oklahoma, Upadhyay spent much of his youth on formative trips traipsing around the Grand Canyon, where his aunt was an artist-in-residence for many years. That's when he first laid eyes on the river—a water body that would end up playing a major role in his career. He went on to earn an economics degree at California State University, Fullerton, followed by a master's in public administration from the University of La Verne outside Los Angeles. Fairly early in his career he joined the Metropolitan Water District of Southern California and over time ended up as the water resources manager—a huge responsibility. The size of Metropolitan's service area is astounding: five thousand two hundred square miles. If the service area were a state, it would be the fifth largest—by population—in the United States. Upadhyay was responsible for ensuring that nineteen million Southern Californians didn't run out of water.

Since the 1940s Metropolitan has siphoned more than a billion gallons per day from the Colorado River. That turned out not to be enough, so in the 1970s it started siphoning millions more from Northern California via the State Water Project. The ecological costs of these huge water diversions made Metropolitan a frequent target of environmental groups. Robbing water from one ecosystem so that people in another ecosystem can drink it (and water their lawns, and fill their pools, and grow crops ...) doesn't go over well with the green crowd. It also takes a lot of energy to move that water. The State Water Project's pumps are the single largest consumer of electricity in California,[1] which layers a colossal carbon footprint on top of the ecosystem robbery tag. Roughly half of the water from the state project goes to Metropolitan.

But in the early 2000s, as Metropolitan's service area entered the worst drought in twelve hundred years, those two big water diversions started shrinking at alarming rates. Projections from Upadhyay's team showed that if they didn't find new sources, "the need for water in Southern California will outstrip our supplies," he told me. "The climate issues, and the drought issues are bigger than they've ever been." Metropolitan, which is based in downtown LA, started looking even more at local water alternatives to offset the declining imports, including conservation, groundwater cleanup, and water reuse.

So it was a man-bites-dog moment in 2008 when Metropolitan—the big water importer—started to think big about water recycling. It reached out to the Los Angeles County Sanitation Districts to talk about a potable reuse partnership. The Sanitation Districts consist of twenty-four separate districts in the county that long ago decided to team up into one combined sanitation program for wastewater and solid waste. They jointly own one of the largest sewage treatment plants in the state, and Metropolitan owns an enormous water distribution network. In 2017 the partners announced plans to build a $14 million potable reuse demonstration facility that would "generate information needed for a potential large-scale regional recycled water program."[2]

That's when it became clear just how big Metropolitan was thinking. The demo plant was small, but the press release referenced an eventual full-scale facility that was huge, producing 150 million gallons of purified effluent per day. Thanks to the new partnership, potable reuse was back on top in Los Angeles, and it was changing Metropolitan's brand. "The big difference with what we are doing," Upadhyay said, "is it would be the first *local* supply that we would own and operate. At that point, you're no longer just an imported-water supplier." Two years after the demo plant was built, Metropolitan and the Sanitation Districts started sinking millions of dollars into planning for the large full-scale facility, which they named Pure Water Southern California—or Pure Water SoCal for short. By 2017 Upadhyay had been promoted to chief operating officer, and his big reuse project was moving quickly—and understandably so. The immense reservoirs on the Colorado River were plummeting, and deliveries from the State Water Project were coming in as low as one-tenth of normal.

Then something unexpected happened: environmental groups and nongovernmental organizations reached out to Metropolitan to help. Although they loathed the water diversions that had been a Metropolitan staple for decades, they were willing to overlook that dark side to support Upadhyay's new reuse initiative. "Now NGOs that philosophically are not supportive of imported water supplies," he said, "are working with us, saying, 'I don't want to talk about the State Water Project or the Colorado River, but I want you to be successful in the Pure Water Southern California program.'"

That optimistic water reuse spirit spilled over to Upadhyay's staff too. With nearly two thousand employees, he can see a hundred and fifty positions turn over in any given year. What's been notable about the new recruits? "Everybody wants to work on our Pure program," he told me. They want to "create a sustainable future, and the Pure Water program—and water reuse at large—is part of that sustainability ethic."

The Pure program is so large that it's changing the way people think about how water is managed on the Colorado River. Upadhyay has been in conversations with his counterparts in Nevada and Arizona about partnering on the initiative. The talks led the water agencies in those states to contribute up to $6 million each in support of environmental planning for Pure Water SoCal. Why? Because if the Pure program is successful, Metropolitan could give up some of its Colorado River water as a trade-off for financial support from the other states. "We can do an exchange," Upadhyay told me. "We might back off [the river] more, and it goes into an account in the name of Southern Nevada or Arizona."

⁓

Upadhyay's main partner is the Los Angeles County Sanitation Districts. Martha Tremblay works as the districts' second in command.

Tremblay was born in Los Angeles County to parents who immigrated from Mexico, neither of whom had an education beyond grade school. She was such a promising student that she ended up at the University of Southern California on scholarships. Then she earned a master's in civil and environmental engineering from the University of California, Berkeley. "I really appreciate my parents coming to the land of opportunity," she told me, "which for me, it definitely has been."

She joined the Sanitation Districts straight out of grad school. Within a few years, Tremblay was overseeing the Montebello Forebay groundwater recharge program near Whittier Narrows. Although the Sanitary Districts are best known for that potable reuse initiative, they have a long-standing purple pipe program as well. But increasingly purple pipe has been losing favor in the LA metro area. It's not seen as the highest and best use of purified sewage, Tremblay told me. Officials want to reserve it for drinking water instead. Tremblay's team has spurned several sewage suitors—reserving the effluent from their huge wastewater plant for Metropolitan. "This is our last untapped source of water," she told me. "And with it being so much water, we need a strong partner to work with, and that's what we have with Metropolitan."

Given how the historic drought has resulted in dramatic cutbacks from the State Water Project and the Colorado River, I asked her if the LA metro area could survive without major investments in water reuse. "No," she said. "It has to happen—and at a greater scale than it's already happening."

Even so, it will be an expensive engineering challenge. The program is estimated to cost $5 billion or more.[3] A key expense will be getting the purified sewage where it needs to go. Most wastewater systems take advantage of gravity to deliver sewage to their treatment plants. So, once a toilet is flushed, that water runs downstream all the way to Tremblay's wastewater facility in Carson. That means that her vast plant lies downhill from much of her service area. But the demand for the purified sewage—and the access point for Metropolitan's expansive distribution system—is uphill. Fulfilling that demand will require up to sixty miles of piping to pump the recycled water to Metropolitan's drinking water plant, where it can be added to the wholesaler's supply network. As the water makes its way to the headworks, some of it will be pulled off to be released into spreading basins en route. The plan is to build the Pure plant in stages, starting with perhaps 30 million gallons of capacity and then working gradually up to 150 million gallons per day by 2032.

Recharging groundwater basins with indirect potable reuse will be the priority—at least a first, Tremblay told me. But direct potable reuse is planned for the Pure Water SoCal program as well. Direct potable reuse has never been allowed in California before, but new strict direct potable reuse regulations were released by the state in 2023, finally opening the door for the most challenging and controversial form of water recycling. "Groundwater recharge is the primary target," Tremblay said, but in coming years, the plan is to send direct potable reuse all the way up to the top of Metropolitan's distribution network. That means that purified sewage from Los Angeles County can eventually be shipped throughout much of Southern California.

Tremblay's partnership with Metropolitan will use the latest technology designed to turn wastewater effluent into drinking water. But as this book went to press, new questions were emerging about the nation's oldest potable groundwater recharge program at Whittier Narrows. Why? PFAS. Remember that Whittier Narrows's water recycling program started in 1962, and unlike Orange County, it does not treat the water with reverse osmosis before sending it to the spreading grounds. (Reverse osmosis removes PFAS to government-mandated levels.) Instead, Whittier Narrows treats the water to tertiary standards—clean enough to swim in, but not clean enough to drink. Then the water is added to the spreading grounds, where the percolation process and residence time in the aquifer make it potable. As mentioned earlier, one could consider the Whittier Narrows program to be water recycling light because it does not use reverse osmosis or some equivalent technology to purify the water before it goes into the ground. But Tremblay and others assured me that decades of regular testing showed that the water has repeatedly met safe drinking water standards.

But in the PFAS era, that seems to be changing. Water recycling light may not cut it anymore. "We do find PFAS in our recycled water," Tremblay told me. "[It] is right at that level of concern." She said that her team is working with the local Water Replenishment District on potential solutions, including additional treatments like reverse osmosis, or some equivalent purification process. "Some wells are affected, some are not affected," she said, adding that at least a portion of the flow is now being purified. But she said more improvements may be required, potentially impacting the entire Whittier Narrows operation. "If it's not PFAS," she told me, "it might be another compound that's going to put us there to provide additional treatment."

⌐

Two years after Metropolitan announced the Pure Water SoCal program, Los Angeles mayor Eric Garcetti declared that his city planned to recycle 100 percent of its sewage by 2035. At more than 200 million gallons per day, Garcetti's project will be at least 50 million gallons larger than Metropolitan's. Hubertus Cox was working for the city's stormwater capture initiative when Garcetti held his historic press conference. Cox had no idea that the announcement was coming. "I think it took some people by surprise," he said. "It definitely took me by surprise."

Six months later Cox was tasked with transforming Hyperion into the largest potable reuse facility in the United States. "We were given the deadline of 2035," he told me. "I see that as being around the corner." Cox is a modest, soft-spoken Dutch engineer. He came to California for postdoctoral research, met his wife, and stayed. Now he has one of the most demanding engineering jobs in the country. "The amount of recycled water that we can produce," Cox said, "will be a game-changer for how we manage water in the City of LA."

The biggest hurdle is morphing the giant Hyperion wastewater plant while it continues to operate. "That is going to be a major challenge," Cox confessed, "and we have to make sure that every single day our discharges to Santa Monica Bay meet all the permit requirements." I told him that the job sounded like upgrading a train while riding it. "That is a fairly good way of saying it," he replied, "and that is why some of our engineers are a little bit nervous."

How will they do it? Just like Metropolitan's Pure Water SoCal project: in phases. The first phase will likely be around fifty million gallons per day. Cox's team will construct the potable reuse operation first, and once it is ready to purify the effluent, one section of the traditional sewage treatment process will be connected to the new water conveyance system. The first batch of recycled water is expected to go into local aquifers—the Orange County method of indirect potable reuse.

Cox estimates that it will take three phases over numerous years before he finally ends up purifying 100 percent of Hyperion's sewage. Thanks to the constant drumbeat about drought in the Southwest, the pressure is ever-present. "I'm trying to figure out how to do it as fast as possible," he told me, "and other people are telling me to do it faster." He arrives at the office at 6 a.m. daily and regularly clocks sixty-hour weeks. "I enjoy every second of it," he said. He's hiring people at a rapid pace while also mapping out the project timeline and meeting with partners, regulators, and a state-mandated advisory panel.

Hyperion's campus covers 144 acres, and there's no room to grow. It is surrounded by four immovable objects: Los Angeles International airport (LAX) on one side, a power plant on another, a residential neighborhood on the third, and the ocean on the fourth. That's forcing Cox to shoehorn his new equipment into a tight space. "Piece by piece," he told me, "we need to take something out and replace it with something else." In addition to the long-term construction planning, Cox is also building a demonstration facility that will send recycled water to LAX for nonpotable uses, as a proof-of-concept exercise. The demonstration building will serve as an outreach venue for public tours as well. He has also erected a membrane bioreactor pilot program on site that is being vetted by the state-mandated independent advisory panel. The bioreactor will serve as the first stage of the reverse osmosis–based treatment process—which is the same overall treatment system that Pure Water SoCal plans to use as well.[4]

After Cox spent two years on the job, a major accident occurred at Hyperion. On July 11, 2021, the weekend crew found itself battling an unusually large surge of debris in the sewage stream that constantly flows into the plant. Like most large wastewater facilities, Hyperion is equipped with a series of bar screens that sift the largest debris from the sewage stream so that it can automatically be raked off and discarded. But on that Sunday afternoon, the detritus could not be removed fast enough. One grate after another became clogged and shut down.

Like a scene from *The Sorcerer's Apprentice*,[5] employees were overwhelmed as the rank liquid spilled into the plant. Staff had problems lifting gates that would have allowed the sewage to bypass the clogged bar screens. As water pressure built up in the catacomb-like infrastructure underneath the plant, lids popped off manholes in the headworks building, which continued filling with sewage. Employees fled for their safety. The putrid water kept coming, cascading throughout much of Hyperion's campus. With no options left, officials sent millions of gallons of raw sewage rushing into Santa Monica Bay. It was one of the worst spills in years at Hyperion, damaging large swaths of the plant's infrastructure, not to mention its reputation.

When Cox came into work early on Monday morning, he learned

that there had been an accident the day before. "I ran to the site," he told me, and that's when he discovered that thousands of gallons of sewage had streamed into his project area. As he approached a large hole that had been dug for the foundation of his demonstration facility, he saw that it "was filled to the rim." A full-sized front-end loader was in the bottom of the pit. "It was submerged to the roof," he said. Once the sewage was pumped out, some sixty dump trucks of soil were removed to decontaminate his entire work site. It was a disaster for Hyperion, for Santa Monica Bay, and for the neighborhood, which put up with nauseating odors for weeks as the disabled plant was repaired. It was a setback for Cox's water recycling operation too. But Cox isn't complaining—he said that he felt terribly for his Hyperion colleagues, especially Timeyin Dafeta, the plant manager.

The crisis prompted an obvious question: if Hyperion can't handle its day-to-day operations, how can it confidently be transformed into the largest water reuse facility in the United States? Cox assured me that if there was another spill—which he described as a once-in-a-lifetime event—his reuse operation would be safe. It will be designed in such a way that sewage "will not get into the recycled water," he said. "So I'm not worried from a public health perspective." But he admitted that the accident could damage the image of his program—or even the reuse industry. "It may have a negative impact on water recycling," he said, "but I think we can make this an opportunity." Cox said that the spill should be leveraged into new investments that upgrade the entire Hyperion campus into "the most modern facility you can find."

Traci Minamide agrees. As the chief operating officer of the city's sanitation department, she serves as Cox's boss and oversees a billion-dollar operation with thirty-five hundred employees. She told me that it has been nearly thirty years since Hyperion received a major upgrade, and the spill has "resulted in a huge capital improvement program." She said, "We are seeing this as an opportunity to say, 'Look, if we need to upgrade the basics at the plant, let's upgrade in a way that keeps in mind the need for that advanced treatment.' We've seen that in multiple places.'"

That makes sense. It's too bad that it took a major spill to make it happen. But a skeptical public is always worried about human error, and the Hyperion accident was an embarrassment to the wastewater industry. The public has zero tolerance for spills that discharge millions of gallons of raw sewage into the ocean. But it's also a reminder that potable water recycling programs—in Los Angeles or anywhere else—need to be flawless as the industry continues to distance itself from the doubts that ran rampant during the toilet-to-tap era. The reuse community has carefully and methodically built back public confidence—not everywhere—but in key segments of the country. But that credibility rests on a wafer-thin foundation that could crack at any time, especially given the long and bitter negative publicity that has dogged potable water recycling since the 1990s. As the industry continues to enjoy a boom in acceptance in places like Southern California, water recycling officials everywhere need to be reminded that there is simply no margin for error. Numerous sources for this book told me that if a reuse health scare were to occur, it could setback potable water recycling again—for a generation.

Minamide hopes that doesn't happen. In her view, the Hyperion reuse project is essential for LA's future. "This is like the new frontier," she told me. "This is where we need to go in order to ensure that we have the water supply that we need."

I asked her if the LA metro area could survive without potable reuse.

"I don't think so—I don't," she said. It's clear that she shares the same pressure that haunts Cox. "Those of us in the industry know why there's this urgency," she told me. "Even moving as fast as we can, is it fast enough?"

⌣

Cox rebounded from the sewage spill, which only knocked him slightly off schedule. From the beginning, his plan has been to build a potable reuse project that would deliver purified water to Hyperion's fence line. From there, the city's water department would be responsible for distributing it to millions of Angelenos—no small feat. Anselmo Collins is responsible for making that happen. A dynamic and charismatic Panamanian immigrant, he leads the city's water division at the Los Angeles Department of Water and Power. Collins is clearly enthused by the challenge and has no problem thinking big about water. "Growing up in Panama, I got to see the Panama Canal," he told me. "The idea of hydraulics was quite interesting to me—that you could do so much with water—and be able to basically cut through a country and connect the oceans."

Cox's purification program is known as Hyperion 2035. But as soon as the water crosses his fence line, it becomes Operation NEXT, the water department's name for the distribution effort. To outsiders, that is a worrisome sign—two departments in the same city using two different names for the same water recycling initiative. To some, the double branding suggests siloed agencies struggling for synergy. Nothing could be further from the truth, Collins told me, adding that the city plans to rebrand the entire program for just that reason. When I asked Collins to describe, for a national audience, just how big the Hyperion 2035/Operation NEXT program was, he said, "It is the biggest water recycling program in the entire nation.... This is truly a transformational project that is going to completely change the water supply in the region."

It will also be expensive. Hyperion's share of the cost is weighing in at around $5 billion. But Operation NEXT's distribution system is the real expense—and may require tunneling under the city to send the recycled water back uphill to tap into the top of LA's water distribution network. That will add another $12 billion or so, for a whopping total of $17 billion—a number that is sure to grow. Pick an inflation rate of, say, 3 percent. Doing a back-of-the-envelope calculation, you can tack on another $510 million *annually* in inflation to the cost of the project, a project that will take more than a decade to complete. "The longer you postpone things, the more expensive they get," one official told me. "We should expedite this and make this urgent. It will benefit us."

Sewer rates in Los Angeles are bound to rise. Water rates too. But let's face it, there are costs to living in an arid climate during the worst drought in twelve hundred years. Yet there's no way that a project of this magnitude will ever be completed without large infusions of state and federal funds. Collins speaks passionately about the need for federal dollars. "Seventeen billion is a solid number, but it's subject to change," he told me. "It is a lot of money, and we recognize that we need to be as aggressive as possible in going after funding." The magnitude of the expense keeps him up at night. "I want to make sure that the cost is not 100 percent passed on to the ratepayers. It's just unsustainable," he said. "A large percentage of our population lives below the federal poverty level."

What's the most promising way to attract federal dollars? Think even bigger, Collins said. How? By combining the city's gargantuan water recycling effort with Metropolitan's massive Pure Water SoCal program. Rather than have the two enormous programs competing against each other for money, Collins said, they should team up and travel to Washington (and Sacramento) together. If they can speak with one cooperative voice on behalf of water security throughout the LA metro area—and beyond—they will have a much better chance of securing the money that they need.

For that argument to be truly convincing, Collins said, the two reuse programs need to be physically connected so that LA's recycled water can be sent to Metropolitan and vice versa. Collins said that he envisions numerous connections throughout their respective water supply networks. "It is certainly a big challenge," he told me, "but the reason

I'm very optimistic is we have a great working relationship with the current general manager at Metropolitan—and the staff." The leadership teams from the city, Metropolitan, and the sanitation districts gather monthly in what Collins calls regional recycled water coordination meetings. "The goal is to find a way to integrate these two huge programs so we can maximize the benefit for the entire region," he said. The uber-partnership would allow LA's recycled water to tap into the backbone of Metropolitan's distribution system, "and that could be another way for us to save money—from an infrastructure and construction perspective," he added.

There are long-overlooked seismic benefits to the program as well. "The three major straws that bring water to the LA area are the LA Aqueduct, the State Water Project, and the Colorado River Aqueduct," Collins told me. "They all cross the San Andreas Fault. So, it's not a matter of *if* we're going to have a major quake, it's when." He said that Operation NEXT will provide "the ability to make us more resilient … and have water—even after a quake."

Almost everyone I spoke to in Los Angeles said that there is no way the city is going to be able to recycle 100 percent of its water by 2035. It's a great stretch goal, but it seems almost impossible to meet, they told me. Most suggested that 2045 was a more realistic target date for Garcetti's original vision. Collins said that he expects some recycled water to be delivered by 2035, just not 100 percent. "By 2035, my expectation is that we're going to have portions of the program done that would allow us to have some early deliveries," he said. "But the entire program … will take at least ten years after that."

What will happen with the water after it leaves Hyperion's fence line? Surprisingly, at full build-out, Collins said that he sees most of that water being delivered as direct potable reuse with no environmental buffer—although the water will be blended with other system water before it is distributed to customers. The rest of Hyperion's water would be banked in the region's bountiful groundwater systems in the Central Basin, West Coast Basin, and San Fernando Basin. Collins and his team see the local Water Replenishment District as a "critical partner" in the groundwater storage effort. "I think direct potable reuse is the way to go," Collins told me. "It's going to give us the biggest benefit, it's going to make us efficient with the current system that we have, and we'll be able to maximize the use of that water for all applications."

Once reached, that goal will put Los Angeles among some elite company in the global water recycling community. Windhoek, Namibia, was the first city in the world to add direct potable reuse to its public water supply, back in 1968. The arid capital was alone in that category for decades. In the 2000s Windhoek was followed by the Texas troika of Big Spring, Wichita Falls, and (someday) El Paso. But none of those programs is anywhere near as large as the combined direct potable reuse vision that is planned by the City of Los Angeles and Metropolitan's Pure Water SoCal project. Together, the two huge California programs eventually plan to deliver direct potable reuse water to millions of people throughout much of Southern California. That will be more than Windhoek, and the Texas troika combined, making it the latest example of how the Los Angeles metro area plans to go all in with potable water recycling.

⌒

The magnitude of the push for potable water reuse in the LA metro area is mind-blowing. In a nation where most people still have no concept that sewage can be safely purified into drinking water, all the key water agencies in Los Angeles are rushing to adopt potable reuse as a major pillar of the regional supply. The metro area that relied on controversial and environmentally damaging water diversions to survive the last century now sees similarly substantial investments in potable water recycling as the secret to surviving in this century. An influential cadre of potable reuse true-believers has coalesced into the right positions at the right time—throughout the metro area—to make that giant vision achievable. That includes Adel Hagekhalil, general manager of Metropolitan, who is a strong water reuse proponent. He helped Mayor Garcetti envision Hyperion 2035 before Hagekhalil left the city for the top job at Metropolitan. The collective excitement among regional water staff is palpable. So are their haunting fears of drought.

Garcetti shares those fears. That's why he launched Hyperion 2035. "I was raised with a drought in the 1970s," he told me. "I remember the first time people said, 'Don't flush the toilet.'" The self-described "water geek" has heard staff concerns about completing the city's water recycling program on time. Fine, he said, hoping that the drought holds off long enough to provide another decade of wiggle room. "The history of this city's growth has been all about water, and the future history of its survival relies on water," he told me. "Without the water, it stops."

In December 2022 the Hyperion 2035 program faced what could have been a loss of momentum. That's when Garcetti, restricted from running again because of term limits, left office. Former LA congresswoman Karen Bass replaced him. Bass had a track record of supporting water reuse, but the mandate of her election came with other priorities, including homelessness and crime. Garcetti's departure is a reminder that the city's historic water recycling initiative could struggle to remain top of mind in the mayor's office as the multidecade project cycles through one administration after another.

But Bass's office denied that there had been a loss of momentum during the transition. Nancy Sutley, Bass's deputy mayor of energy and sustainability, agreed that the new mayor had been elected on a housing and crime mandate, but that does not mean that the city has pulled back on potable reuse. "Everybody's moving full speed ahead," she told me. "I don't think there's a loss of momentum, and there's a lot of work to do."

Climate change has challenged Garcetti's hometown to step up. From the start, he knew the vision would have to transcend mayoral administrations—it's too big of a project not to. But it's not about one mayor; rather, it's about the need for the nation's largest arid city to survive the climate era. "We just have to have the political will," Garcetti told me, "to work out the engineering feat of this generation." Other cities do too. "I don't know why this wouldn't become standard operating procedure for the entire Sun Belt," he said. "It makes some sense even in less water-stressed places.... It makes a lot of sense to reuse what we have. It's a central tenet of sustainability."

Whether Hyperion meets the 100 percent goal in 2035—or 2045— Garcetti is thrilled to see the LA metro area focusing so collectively on the region's long-term water supply security. "The history of water policy is always ignore, ignore, ignore—panic, quick fix—ignore, ignore, ignore—panic, quick fix," he told me, "as opposed to focus, focus, focus, long-term solutions, no panic, no stress—it's a better way to live." The City of Angels is trying to position itself for that happy ending. A generation ago, as the metro area struggled through the dark depths of the toilet-to-tap era, few would have predicted that potable water recycling would one day be seen as the city's savior. But we're in a new water generation now. Call it the potable reuse era. As climate change continues to bear down on the public water supply in so many places, it will be fascinating to see how many other communities—in California, or the nation, or the world—will follow LA's bold, ambitious, and courageous water recycling lead.

CHAPTER 12

Water Diversion, or Water Reuse?

Central and western United States

THERE'S NOTHING QUITE LIKE A ROOM FULL of a thousand water-stressed people. Thanks to the worst drought in twelve hundred years,[1] the Colorado River Water Users Association annual meeting has become a high-stress affair. Water officials, farmers, scientists, and tribal members gather annually in Las Vegas, and every year the meeting seems to grow more intense. "I can feel the anxiety—the uncertainty—in this room," said Camille C. Touton, commissioner of the US Bureau of Reclamation in 2022. "The hotter and drier conditions that we face today are not temporary."[2] Her remarks came just a few weeks after the one-hundredth anniversary of the Colorado River Compact,[3] but the mood was far from celebratory. Yes, the river has seen a few wet years since the turn of the century,[4] including 2011, 2017, 2019—and the atmospheric rivers of 2023. But the overall trend is clear: the watershed is getting drier.[5] Since 2000 the river has slipped deeper and deeper into a climate-driven contraction that threatens the welfare of forty million people from Wyoming to northern Mexico.

The annual Colorado River conference features panel discussions like "Trade-offs and Turbulence," "Adapting to the New Normal," and "Colorado River 101." But for years the takeaways have remained the same: (1) there is not enough water in the river; (2) climate change has altered the watershed's hydrology in new and alarming ways; (3) as the multilayered, incremental water agreements overseeing the river grow more complex, they have become increasingly inadequate; and (4) perhaps most importantly, no one ever really has a permanent solution. As a result, the river is always the biggest loser. "Ultimately, I think what we're going to see here is some major rewriting of Western water law," predicts Brad Udall, a leading expert on the river from Colorado State University. "How this works out is anybody's guess."[6]

One of the most memorable panels from the 2022 conference was titled "Evaluating Collaborative Options for a Second Century." It was moderated by Terry Goddard, former Phoenix mayor and Arizona's attorney general in the early 2000s.[7] Goddard now sits on the board of the Central Arizona Project, or the CAP, as it's known—the 336-mile federally subsidized[8] water diversion that siphons up to 456 billion gallons of water from the Colorado River annually.[9] The CAP "is an engineering marvel," its website says, "that has contributed dramatically to our quality of life and the sustainability of the state's water supply and economy."[10] That said, the website also notes that more than twenty-four billion gallons is lost to evaporation and ground-seepage before reaching its destination.[11] It was the last major water diversion built in the United States.

So it was fitting that Goddard's first speaker was pitching another far-flung water diversion, this one being much more ambitious than anything that had ever been built in North America. It was so ambitious that it bordered on folly. The presenter was Mark Rude, representing the Southwest Kansas Water Management District Number Three. He was also representing the Kansas Aqueduct Coalition, which is based in Alma, population eight hundred. He was pitching a mammoth water project that would pull millions of gallons from the Missouri River—at the northeast corner of Kansas—and pump it across his state. Along the way, the water could be used for irrigation and, more importantly, to recharge the depleted Ogallala Aquifer, which has plunged by more than one hundred feet in some areas.[12] Once the Missouri's water crossed Kansas, Rude said, it could continue on up over the Rocky Mountains and connect to a stream in the Colorado River watershed, sending rescue water to the parched Southwest. He made it sound so easy.

But Rude struggled to produce a convincing pitch. He admitted that the program might create "potential interstate opposition" and even "potential source-state opposition" from his fellow Kansans, not to mention opposition from the people whose land would be taken away to build the canal. He also admitted that there "may be concern regarding costs and financing." Other than that, things were looking good.

But not really. He deftly glossed over a few other downers, including federal and state permitting nightmares and wave after wave of inevitable lawsuits from environmental groups, any one of which could kill the whole thing. There was no mention of what environmental damage might be wrought to the Missouri River or what harm might occur to downstream water users—like St. Louis.

Figure 12-1. The Central Arizona Project diverts water from the Colorado River for irrigation and municipal use in parts of central and southern Arizona. (US Bureau of Reclamation)

The kicker came when Rude admitted that his coalition couldn't even afford to get the water from one end of Kansas to the other—not to mention shipping it all the way to the Southwest. A key expense would be the sixteen giant lift stations required to push the water uphill across Kansas, not to mention the enormous energy costs that are required to move water against gravity. Compared to the other states, Kansas ranks seventh in terms of flatness.[13] One can only imagine how many lift stations it would take—and what the energy costs would be—to raise that water thousands of feet to get over the Rockies.

The more Rude spoke, the more unrealistic the whole thing seemed. As the expression goes, he was all hat and no cattle. That he made it onto the speaker docket at the conference shows how desperate some people in the Colorado River watershed have become. What Rude was really seeking was a Southwest sugar daddy. "We're looking for collaborators," he said, who could help pay for his colossal enterprise. But he was in the wrong town. The sugar daddy doesn't live in the Southwest. He lives in Washington, DC, and his name is Uncle Sam. No program of this sort could ever be built without major assistance from US taxpayers, as was the case with the much smaller (but still very large) $4.4 billion CAP[14] ($9.2 billion in 2023 dollars). The original price tag on the controversial CAP in 1965 was $832,800,000.[15]

After years of litigation,[16] construction on the CAP started in 1973 and was completed in 1993.[17] During the next thirty years Arizona's population doubled, fueled by all that Colorado River water pouring into the desert. But by 2023 the CAP was falling short on water deliveries,[18] thanks to drought, burgeoning population growth, and unsustainable agricultural practices—like growing water-guzzling alfalfa and cotton in the desert.[19] That created a semistranded asset for US taxpayers[20]—they helped fund a larger canal than Arizona is now able to fill. Because the CAP was still strapped with fixed costs, the price of the water in the canal went up as water deliveries declined.[21]

After Rude completed his remarks about the Kansas aquifer idea, Goddard, the moderator from the CAP, asked a salient question: "How do the good people of Kansas feel about moving their water out of state? ... I think everybody here has a concern about that." For the next forty-five seconds, Rude rambled on without ever answering Goddard's excellent question.

Robert Glennon is one of the nation's leading water experts. I reached out to get his reaction to Rude's Missouri River diversion proposal. Glennon, the Morris K. Udall Professor of Law and Public Policy, emeritus, at the University of Arizona, has written three highly regarded books about the water crisis. When I told him about the Kansas pitch, he struggled to contain his dismay. "These proposals are complete pipe dreams—full stop," he told me. "There are so many obstacles to doing that—so many hurdles—there's one particular hurdle: something called the Rocky Mountains." He said that moving Missouri River water up over the Rockies would use "an insane amount of energy."

Then he turned to another quixotic diversion proposal favored by many politicians in his home state of Arizona. That plan proposes building the largest desalination plant in North America—by far—and then piping the water from Mexico nearly two hundred miles uphill to a reservoir west of Phoenix.[22] "Just think about locating a plant of that scale on a pristine beach," Glennon said. "Then you have to have a powerplant equal in size to produce enough energy. That means transmission lines to this very lowly populated area.... It's not even clear where you have to go to get the electricity.... And what exactly is in this for Mexico?"

Then there are the cost estimates for the desalination project. Glennon said that unrealistic figures of $5 billion have been floated, along with claims that the desalination plant and the accompanying water diversion infrastructure can be built in just four years. "Nonsense!"

Glennon told me. "It's going to be $100 billion," and the earliest a giant plant like that could be brought online would be 2040. But in our conversation, he made it clear that the project should never be built. "What have you been smoking?" he asked of the plant's promoters, including former Arizona governor Doug Ducey. "A proposal like that," he told me, "is so ill-conceived on so many levels."

There are many other obvious solutions to the Southwest's water crisis, Glennon said. They include more conservation, especially on agricultural lands, and investments in water recycling.[23] "Half of the irrigation lands in the West are flood-irrigated," he said. "That's just a tremendous waste." What's more, much of that land is used for water-intensive crops, especially cotton and alfalfa. The alfalfa primarily goes to feed dairy cows in the desert.[24] Dairy cows consume an enormous amount of water—up to fifty gallons per animal per day—and when they're heat-stressed, that water consumption can double.[25] That's why some desert dairy farms use showers to cool down their cows, consuming even more water.[26] Amazingly, dairy has become the number one agricultural commodity in Arizona.[27] (Beef cattle are a close second.)[28] That raises an interesting question: should Arizona be importing water through interstate diversions, or should it just truck in milk instead? "Agriculture is going to be the one entity that gives up the most water because [farms] consume 80 percent of the water," Glennon told me, which is why he supports federal and state incentives to help farmers conserve by modernizing farm infrastructure.

As the Southwest drought has worsened, increasing scrutiny has been directed at crops grown with Colorado River water. That's particularly true for crops that are exported. "We spend an enormous amount of water growing alfalfa and Sudan grass and other forage crops and a very high percentage of that is exported to Japan, China, and Saudi Arabia," complained John Entsminger of the Southern Nevada Water Authority. On the other hand, he said, winter fruits and vegetables from the

Southwest are extremely important. "You can't grow lettuce in Iowa when it's zero degrees outside in January."

Maybe not Iowa, but how about Minnesota? That's where my January greens come from.[29] Thanks to a $68 million infusion of capital from a sustainability-focused asset management firm,[30] Minnesota-based Revol Greens is now expanding its greenhouses into California, Georgia, and Texas.[31] Arizona farmers like to say that the United States gets the vast majority of its leafy green vegetables from the Yuma area during the winter.[32] Although that may be true now, the salad market is changing. Wall Street—or at least those financiers who care about sustainability— is starting to bet against the Yuma Valley. With dead pool capturing headlines, who can blame them? I buy hydroponic greens because they are fresh, and usually arrive in stores within twenty-four hours of being picked. They taste better and last longer in the refrigerator. What's more, they don't carry any of the Colorado River baggage.

Rather than dreaming about unsustainable water diversions from the Missouri River or the Sea of Cortez, Glennon said, we should be investing precious infrastructure dollars in recycled water instead. "I think there will be a bright future for potable water recycling," he told me. It is a renewable, local, drought-resistant supply that does not carry anything close to the unsustainable environmental, financial, or energy drawbacks of water diversions. "This is real water, wet water," Glennon said in reference to reuse, "not pipe dreams." What's more, the supply of recycled water grows with the size of the population as people produce more sewage. "There will be more water that ends up in the treatment plants that can be available locally for reuse."

⌐

But the fanciful diversion proposals keep coming—mostly out of Glennon's home state. In 2021 the Arizona legislature formally asked Congress to study diverting floodwaters from the Mississippi River and shipping them to the Southwest.[33] When that didn't happen, Arizona decided to go it alone. "In the West, the long-standing tradition is that when you don't have enough water to serve your city or state, you take it from somewhere else," wrote Tony Davis, a longtime environment reporter with the *Arizona Daily Star*. "With a water crisis staring it in the face, Arizona's Legislature is about to go down that path again—by creating a new fund with a starting cost of $160 million to import more water, possibly from as far as the Mississippi River."[34]

The following year the state set aside $1 billion to explore water supply alternatives, including water diversions.[35] Glennon said that Arizona's desperate legislators are overlooking solutions that are right in their backyard—and their toilet. "You're going to build [a pipeline] for a couple thousand miles across the United States? When you could maybe stop growing grass in backyards? When you can fund farmers to modernize their infrastructure? When you could do a lot more with reuse? To me, the right answer is really obvious," he said.

Then there was the California diversion proposal that went viral in 2022. An unknown retired engineer published a letter to the editor proposing to divert water from the Mississippi River, starting at a spot not far from its terminus in Louisiana.[36] He said that Cajun Country didn't need that water, so no harm done.[37] His back-of-the-envelope calculation showed that—once his project was built—he could allegedly fill Lake Powell in 254 days and Lake Mead in 370 more. All it would take is a huge diversion moving water uphill at the extraordinarily fast clip of 250,000 gallons per second.[38] Problem solved. The letter went viral,[39] which was more a statement of the times we live in than the engineer's expertise. As the media flurry continued, this proposal was first debunked by the *USA Today* network[40] and further debunked by three researchers at Western Illinois University.[41] But in the clickbait era, the engineer kept getting missives published.[42] He was just trying to help, which is good. Raising false hopes is not.

So let's debunk the proposal once and for all. First, the size: at 250,000 gallons per second, he's talking about building a canal that is 50 percent *larger* than the average flow of the Colorado River itself.[43] Then he envisions pumping that water uphill, from near sea level in Louisiana to the Front Range of the Rockies. As the researchers from Western Illinois University said, no such pumps currently exist that can handle such an enormous task.[44] From the Front Range one would presumably need an even more powerful series of pumps to lift the equivalent of one and a half Colorado Rivers thousands of feet over the Continental Divide. The bottom line? Filling Powell is not the challenge. Getting the water there is.

It took twenty years to build the much-less-complicated CAP, and that was after years of legal delays. How long would a Mississippi River project take to build? Maybe forty years? And how long would the legal battle drag on? Tack on another decade? (That's assuming the environmental litigation miraculously proved unsuccessful.) Who knows what will happen to the cities, and the farmers, and the cows, in the desert Southwest by then. What's more, it turns out that Louisianans *do* need the waters of the Big Muddy. Those sediment-laden flows play an important role in rebuilding the rapidly declining Louisiana coastline, which is in a losing battle with sea level rise.[45]

It's hard to imagine that Congress would ever support such a plan. Mississippi River diversion proposals, from the engineer, and others, during 2022, generated hate mail from the river's headwaters to Gulf of Mexico. "Californians should remember their own history, namely the Owens Valley water wars when valley farmers dynamited an aqueduct that was stealing their water and draining it into the sewer that is Los Angeles," wrote one angry river rat. "We have plenty of dynamite in Minnesota."[46] Or as one Missourian put it, "Leave it to the Westerners to come up with solutions to their problems by causing problems for others.... [This] is nothing more than a plan to steal, under federal-government oversight at taxpayers' expense, water that belongs to the

Midwest."[47] And consider this note from Louisiana: "Politicians and residents out West have created this crisis over the past century," one man wrote. "If anyone tries to steal this water, they'll find themselves outvoted in Washington and losing in court." [48]

Wouldn't it make more sense to just recycle sewage throughout the Colorado River watershed instead? It would be exceedingly cheaper, with a fraction of the carbon footprint and nothing close to the environmental drawbacks of water diversions. Nonpotable water recycling is happening throughout much of the Southwest, but several states could do a lot more with potable reuse, especially Arizona.[49] "Everything that hits a drain in Las Vegas is treated, put back in Lake Mead, and can be taken right back out," said John Entsminger at the Southern Nevada Water Authority. "We have 100 percent reuse of our indoor supply, and it's critical that areas like Phoenix, Denver, and Los Angeles do the same thing."

In 2023 Phoenix took Entsminger's advice, announcing a major potable reuse program of its own. For years Arizona's largest city has sent more than fifty million gallons per day of nonpotable recycled water to the local nuclear power plant. But as the Southwest's megadrought has grown more severe, Phoenix now plans to produce up to ninety million gallons per day of direct potable recycled water by 2030. "[With] what we've seen on the Colorado River, there's no time to waste," said Troy Hayes, director of Phoenix's Water Services Department. "We have an obligation to provide safe, clean drinking water to our customers forever and ever," he told me. "In times like this, that's a challenge."

⌣

I have been writing about water diversion schemes since the early 2000s. Proposals to divert water from the Mississippi, or the Missouri, or even the Great Lakes, are nothing new.[50] In 2011, when Pat Mulroy was head of the Southern Nevada Water Authority, she pitched the Mississippi River diversion idea to the US Chamber of Commerce[51] only to see it slide into obscurity. A year later, in a report full of desperate ideas, the US Bureau of Reclamation examined diverting water from the Missouri River—or the Mississippi—as a form of Colorado River relief, but the ideas went nowhere.[52]

Or the most ridiculous of them all was the North American Water and Power Alliance, or NAWAPA. It proposed damming up several large rivers in Alaska and Canada in the 1960s and then sending that water south to large reservoirs that could be created using nuclear weapons.[53] There was also the GRAND Canal proposal of the 1960s, which is sort of a Canadian version of NAWAPA but without the nukes.[54] Both programs advocated for artificially replumbing the North American continent, eviscerating one natural water system after another in the process. Each idea eventually collapsed under its own extraordinary fiscal and environmental weight.

These crazed ideas—and others like them—led to the adoption of the Great Lakes Compact in 2008. The compact bans long-range, large-scale diversions from the lakes, which hold 20 percent of all the fresh surface water on the planet. Great Lakes officials saw the Southwest's water nightmare coming a long time ago, so they adopted legislation to protect the lakes from damaging water diversions. Planning for the compact began in 1998, when Lake Powell and Lake Mead were brimming. But the philosophical origins of the Great Lakes Compact go back to the 1980s.[55] To Great Lakes officials, the latest water diversion schemes vindicate their proactive water management approach. "It's always disturbing to hear these pleas for help from parts of the country that haven't done nearly enough to help themselves," said Todd Ambs, who helped negotiate the Great Lakes Compact and currently chairs the Great Lakes Commission. "Especially when they're talking about projects that are both fiscally and environmentally irresponsible."

Ambs told me that many of those who drafted the historic Great Lakes agreement feel that the Colorado River Compact process has long been broken. He said that southwestern water officials need to make the transformative changes required to modernize the century-old document and its numerous subsequent related agreements. "We spent years—and many, many long hours of work—to get a sustainable water management system in place for the Great Lakes region," Ambs said. "These places have to do the same work. It's just not acceptable for them to say, 'Well, that's too hard. We're going to steal the water from somewhere else.'"

With the Great Lakes protected, western water hawks have set their eyes on the Mississippi and Missouri. Many of these proposals are billed as flood control programs, with the "surplus" floodwaters captured from the Midwest and delivered to the Southwest as drought relief. Few in the heartland believe that in the rare event that a canal were to be built southwestern officials would be satisfied with just the floodwaters. As soon as the next drought headline emerged, with the canal's water diversion infrastructure already in place, they would be clamoring for even more flows. The Mississippi can go more than twenty years without flooding.[56] Imagine if that enormously expensive canal was just sitting there empty for twenty years while the Southwest drought continued to get worse. Officials in Arizona and elsewhere would certainly demand more water. "It's a silly idea," Ambs told me. "You can't fix a water crisis in one part of the country by making a water crisis in another."

Are these diversion proposals even the best political strategy? At the rate things are going, the Southwest will likely need a lot more federal drought relief. There are seven Colorado River states. There are thirty-one Mississippi River states. Do southwestern officials really want to poke the water bear? Proposing unrealistic and controversial diversions that roil half of Congress does not seem like the best way to create a sympathetic audience for more western water funding.

It's enough to raise the ghost of Henry M. "Scoop" Jackson, the renowned senator from Washington State. In the 1960s, when the CAP was being debated on Capitol Hill, there was a lot of talk about diverting water from the Columbia River as well,[57] which carves much of the border between Oregon and Washington. Jackson came up with a brilliant strategy: as chair of the US Senate's influential Interior Committee, he pushed through a ten-year moratorium that prevented the federal government from even studying the idea of diverting water to the Colorado River from other watersheds.[58] Given the number of states in the Mississippi River watershed, you'd think that there would be enough political power to revive Jackson's strategy of old—to protect the Mississippi and Missouri rivers—in the unlikely event that these dubious water diversion proposals ever make it to Capitol Hill.

Every generation or two, usually during a drought, these far-fetched diversion plans are furiously floated, only to fade away once people in power realize how hollow they are. "These ideas to augment Colorado River supplies … are a desperate form of denial," said Peter Gleick, a leading water expert at the Pacific Institute in Oakland, California. He said that they are an attempt to "avoid discussing what everyone knows, deep down, must be done—deep cuts in water use, especially, but not only, in agriculture."

Gleick, who has also written several influential water books, said that the practice of stealing water from one ecosystem and sending it to another needs to stop. "The era of new, large-scale, long-range traditional supply projects, including diversions and dams ought to be over," he told me. "There are a few people, and organizations, and institutions, who haven't realized that it is over." Similar to Glennon, Gleick agreed that the Southwest has several much more sustainable water supply options to help it through the climate change era, including conservation and water recycling. "We can no longer assume the traditional solutions will work," Gleick said, "and we can no longer ignore these nontraditional solutions like water reuse."

Many southwestern water officials agree, including Nancy Sutley, who serves as deputy mayor of energy and sustainability in Los

Angeles. Although her city has depended on long-range water diversions for years, she said that this strategy has become increasingly unreliable under climate change. That's why her city has chosen to invest in water recycling for new sources of supply instead. "We do rely on imported water coming from far away," she told me, "and we see how stressed those systems are. So doing more of the same doesn't seem like the best strategy."

Which brings us back to that panel at the Colorado River conference in 2022. The discussion opened with our man from Kansas and closed with Shanti Rosset from the Metropolitan Water District of Southern California. Metropolitan is widely seen as the leviathan of the water diversion era and has been addicted to long-range water imports from the Colorado River and Northern California for many years. But Rosset had not come to talk about diversions. She was promoting water recycling instead—specifically Metropolitan's Pure Water Southern California project. That profound philosophical U-turn at Metropolitan—pivoting toward reuse—is historically significant in and of itself. Perhaps nothing serves as a better example of how out of favor water diversions have fallen while water reuse has been on the rise. "The program would provide a new purified source of water for Southern California," Rosset said. "This is the first time we are developing an in-region water supply."

Metropolitan's reuse program is unlikely to be litigated because environmentalists resoundingly support it. The program also stands little chance of becoming an unrealistic pipe dream given that a demonstration plant has already been built and the full project is slated for completion by 2032. Rosset reminded the audience that, unlike water diversions, "recycled water is drought-proof and readily available—rain or shine."

When she finished speaking, it felt like we weren't in Kansas anymore.

As Rosset's presentation showed, potable water recycling is booming in Southern California, and it's on the rise in many other parts of the United States as well. A growing number of water thought leaders recognize that wastewater is a vital resource that should not be wasted. As climate change squeezes water supplies in large sections of the country, sewage has emerged as a surprising savior that can bring badly needed relief to water-stressed communities from coast to coast. "Water reuse is the future of water management," predicted Patricia Sinicropi, executive director of the WateReuse Association. "Technology exists that allows us to clean any volume of water, for any beneficial purpose, and at any scale."

The federal government has increasingly embraced water recycling as well. In 2020 the US Environmental Protection Agency (EPA) released the National Water Reuse Action Plan "to accelerate the consideration of water reuse as a tool to help ensure the security, sustainability and resilience of the nation's water resources."[59] Among other things, the WRAP, as it is known, could help streamline the patchwork of state-by-state regulations that currently apply to water recycling programs across the country. It has already helped coordinate a more cohesive reuse strategy throughout the federal government. "It's vital that the federal agencies continued to talk to each other, and to our partners across the water sector, to move the needle forward on reuse," said Sharon Nappier, EPA national program leader for water reuse.[60] Federal funding for reuse is growing as well. The bipartisan infrastructure act of 2021 included $1 billion for water recycling projects.[61]

The water reuse movement is transforming the way millions of Americans think about water in California, Texas, and Nevada—some of the driest states in the country. But it is also changing the way people think about water in "wet" places too, like Florida and Georgia—even New York City. Who knew that the Big Apple had a water recycling grant program?[62] A few high-profile nonpotable projects are up and running in the city, including one at the New School, which collects water from sinks, showers, and washing machines and reuses it in toilets and a cooling tower.[63]

But Manhattan's most famous reuse project is the Solaire Building in Battery Park City, New York's first wastewater recycling program. Its 293-unit apartment complex recycles twenty-five thousand gallons of sewage daily for things like flushing toilets and cooling.[64] Officials are promoting these projects to bring badly needed relief to the city's over-taxed sewer system. In a scene resembling something from the developing world, New York's sewers are so stressed that officials have asked residents to refrain from flushing their toilets when it's raining.[65] That's one embarrassing option. Investment in more on-site water reuse might be a better approach.

Water recycling is happening in other surprising places. Since 2014 Clean Water Services in Hillsborough, Oregon, has provided potable recycled water to home brewers so they can "showcase their brewing creativity" and raise awareness "about the reusable nature of all water."[66] On the other side of the state, the city of John Day plans to utilize recycled water in a local greenhouse as part of a multipronged effort to revitalize the community.[67] Back East, the New England Patriots football stadium even has a water recycling program. A one-million-gallon tank captures wastewater flows during games—so all those fan flushes don't overwhelm the small sewer system in Foxborough, Massachusetts (population seventeen thousand). Long after the game is over, the sewage is recycled for things like irrigation. It's the largest reuse program in the state.[68]

Want your own water recycling system? An Ohio company can help. In 2014 the Ohio Department of Health gave the green light for a firm to start manufacturing the nation's first closed-loop direct potable reuse systems for homes and businesses. These on-site direct potable reuse systems are being produced in a water-rich Great Lakes state. The Western Reserve Land Conservancy in Moreland Hills, outside Cleveland, has installed one of these drinking water systems, creating what some have called the longest running domestic direct potable reuse operation in the United States.[69] In yet another sign of potable recycling's growing acceptance, Colorado adopted direct potable reuse regulations in 2022, followed by California in 2023.

There are plenty of unheralded reuse projects in other areas as well. The famed San Antonio Riverwalk has long depended on recycled water for its flows. Most streamside diners have no sense of how special the water is that flows by their feet.[70] In the Atlanta metro area, both Clayton County[71] and Gwinnett County[72] have successful indirect potable reuse programs that return badly needed treated effluent to regional drinking water reservoirs.

Then there's the innovative water recycling operation at Grassland Dairy outside Eau Claire, Wisconsin. Pinched by insufficient local groundwater supplies, Grassland installed a reverse osmosis–based direct potable reuse system right on the farm.[73] Imagine that: direct potable reuse—the most cutting-edge form of water recycling—in the water-rich state of Wisconsin.

Let's unpack that one for a minute. Wisconsin is bordered by the largest river in the United States to its west (the Mississippi). It's also bordered by the largest inland lake in the country to its east (Lake Michigan). And it is bordered by the largest lake in the world to its north (Lake Superior). It has fifteen thousand inland lakes and more than twelve thousand streams and rivers. The state is literally awash in water. But in Grassland's neighborhood, the groundwater system is surprisingly lacking. Rather than try to divert water from one of the myriad surface water supplies in the state, Grassland's inventive leaders turned to water reuse instead. Is there a lesson there for the desert dairies of Arizona? Or, if not for them, perhaps for their water-groping legislators who are obsessed with looking outward, instead of inward, to solve their predictable water woes?

Can water recycling save the Southwest? That depends on who you ask. John Entsminger from the Southern Nevada Water Authority said that reuse has a critical role to play in southwestern cities. But with 80 percent of water going to farms in the lower Colorado River watershed—most of which are far from sewage supplies in large urban areas—the technology cannot serve as a big-picture solution for the region. "Water recycling is a critical component for municipal users," he told me, "but it's not going to do anything at all to solve the agricultural problems." Deven Upadhyay with the Metropolitan Water District is not so sure. He pointed to Israel as an example, noting that the arid nation uses desalination for drinking water and water recycling to irrigate agricultural lands. Perhaps no country in the world uses water recycling so intensively for agriculture. If things got bad enough, Upadhyay said, "you could see a design like that [in the Southwest]," he told me. "It would be difficult to do, but you wouldn't take that off the table far down in the future." Given the region's water uncertainties, Upadhyay said that southwestern farmers may be prompted to embrace water recycling more than ever. "If you get to a point where ag folks are like, 'My gosh, I simply don't have any water,' to the extent that there's an ability to take recycled water and apply it to ag, they would do it," he said.

But in urban areas, "the big enchilada is direct potable reuse," said G. Tracy Mehan III, executive director for governmental affairs at the American Water Works Association (AWWA). The AWWA calls itself the largest and oldest water association in the world. With more than fifty thousand members, its leadership team knows a thing or two about water. Mehan said that he sees direct potable reuse as a tremendous opportunity—and a challenge. If direct potable reuse can truly deliver safe, reliable water to major metro areas, "it will go a long way toward relieving water shortage problems," he said. But it will have to prove itself. "We have not done this at scale yet," he cautioned. "An unfortunate accident, because of skipping essential steps, would really be a huge setback."

The water reuse industry is worried about that too. The concerns are not focused on the large direct potable reuse programs, like the one proposed for Los Angeles. Rather, the unease lies with systems in smaller communities that may not have the capacity and training needed to operate highly sophisticated water purification facilities. Several people interviewed for this book mentioned Cloudcroft, New Mexico (population eight hundred), as an example. Cloudcroft is widely known in the water reuse industry for having struggled to install a successful direct potable reuse system.[74] "We can't afford a Flint, Michigan, kind of episode—we simply can't afford it," warned Upadhyay. "It could have a public perception impact for all of us.... There's been a lot of discussion about the technical, managerial, and financial competence of water agencies—particularly smaller agencies."

Experts assure us that the technology is safe, but will they continue to get buy-in from the public? Tampa's struggles offer an acute reminder that water reuse proposals can fail without stellar communication and outreach programs, especially in the face of articulate and organized critics.

The water reuse movement has certainly been through some dark days, especially in San Diego, Los Angeles, and Tampa. But many now believe that the toilet-to-tap era is fading, at least in key sections of the country. "People are realizing the preciousness of the resource and that we really should be using it more than once, and we can," said Jennifer West, managing director of WateReuse California. Added Gleick from the Pacific Institute, "The old fear and ignorance about what 'water reuse' meant is slowly being replaced by the realization that, in some ways, this is the best water in the world." Gleick said that recycled water "is moving from being perceived as something bad, to being perceived as something not just good, but urgent. We need it." Felicia Marcus, former chair of the California State Water Resources Control Board, agreed. "Between conservation and water recycling," she told me, "I think we can drought-proof our major cities for decades to come."

With a prognosis like that, who needs water from the Mississippi?

Epilogue

HAVE YOU EVER BEEN TO THE MOUTH OF THE COLORADO RIVER? It's a sad and depressing place—especially if you've read *A Sand County Almanac* by Aldo Leopold. Leopold grew up on the Mississippi River in Burlington, Iowa. In the early 1900s he went off to the Yale School of Forestry where he was mentored by the legendary Gifford Pinchot.[1] As a young forester, Leopold was schooled again—this time on horseback—in the remote national forests of Arizona and New Mexico. There he fell in love with the Southwest and helped create the nation's first federally designated Wilderness Area on the Gila National Forest in 1924. He became one of the United States' most celebrated conservationists, largely as a result of his seminal book, which was published posthumously in 1949. It has been translated into sixteen languages, and more than two million copies have been sold.[2]

But in 1922—the year the Colorado River Compact was signed—Leopold paddled the river's remote, lush delta with his brother. Decades later he included several precious passages about the journey in his book. The chapter referencing the delta's "green lagoons" and "their deep emerald hue" is particularly famous. It paints a picture of the vibrant, fertile ecosystem that enveloped the meandering Colorado as it slowed down and fanned out into a series of forested fingers in the last several miles before the river reached the Sea of Cortez. "On the map the delta was bisected by the river, but in fact the river was nowhere and everywhere, for he could not decide which of a hundred green lagoons offered the most pleasant and least speedy path to the gulf," Leopold wrote. "So he traveled them all, and so did we. He divided and rejoined, he twisted and turned, he meandered in awesome jungles, he all but ran in circles, he dallied with lovely groves, he got lost and was glad of it, and so were we."

"Dawn on the delta was whistled in by Gamble quail, which roosted in the mesquites overhanging camp…," Leopold continued. "A verdant wall of mesquite and willow separated the channel from the thorny desert beyond. At each bend we saw egrets standing in the pools ahead, each white statue matched by its white reflection…." Leopold was clearly moved by the landscape's water-based abundance. "The origin of all this opulence was not far to seek. Every mesquite and every tornillo was loaded with pods. The dried-up mudflats bore an annual grass, the grain-like seeds of which could be scooped up by the cupful. There were great patches of a legume resembling coffeeweed; if you walked through these, your pockets filled up with shelled beans."

When published, these passages read like poetry. Today they read like a eulogy. The fertile Eden that Leopold describes is gone. In 1928 Congress passed the Boulder Canyon Project Act, authorizing the construction of Hoover Dam and the All-American Canal, which diverts much of the Colorado's flow—just as the river approaches the Mexican border. The canal sends that water to California's immense Imperial Irrigation District, which by itself has claim to about as much of the Colorado River's water as the states of Arizona and Nevada combined.

The dam and the canal helped choke off the waters that fueled the delta's teeming ecosystem. Years later, when Leopold sat down to write about his canoe trip, he knew the river had been changed forever. "By this time the delta has probably been made safe for cows, and forever dull for adventuring hunters...," he wrote. "I am told the green lagoons now raise cantaloupes. If so, they should not lack flavor." Tellingly, he had no desire to see the delta again—or what, if anything, was left of its thriving ecosystem. "To return not only spoils a trip, but tarnishes a memory," he said. "For this reason, I have never gone back to the delta of the Colorado since my brother and I explored it, by canoe, in 1922."[3]

I *have* gone back to the delta. Thanks to the many diversions that exist upstream, the waters of the Colorado now rarely reach the Sea of Cortez, making the delta a dry and desolate place. Thanks to the hard work of some environmental groups, small pockets of restoration have emerged from the desert along the river channel. But these important projects represent a mere fraction of the lush ecosystem Leopold explored. Throughout much of the delta, the native flora has been widely displaced by invasive, water-intensive salt cedar. The only sign of wildlife I saw in the lower delta was the track of a sidewinder rattlesnake zigzagging across the barren sand.

What has happened to the Colorado River's delta is a stinging indictment of the damage water diversions can do to a once-thriving river. Diversions are an expensive, environmentally harmful, blunt instrument. They are the last century's way of dealing with water scarcity. Conservation and water reuse are key solutions of the future. Today we have options that are cheaper, more sustainable, and less damaging than water diversions. As Professor Glennon said, the answer is obvious.

No one has ever ruined a river by recycling water.

Acknowledgments

MANY WHO WORK IN THE WATER RECYCLING INDUSTRY have been unfairly maligned for years. Some even show signs of what we might call toilet-to-tap PTSD. So I would like to start by thanking the often-hesitant water recycling practitioners who had the courage to tell me their stories. Most are quoted in the preceding pages, so I won't name them here, but there's no way this book could have been completed without their help. Several early conversations with water experts were particularly influential, including those with Peter Gleick, Mike Lee, Peter MacLaggan, Felicia Marcus, David Sedlak, Patricia Sinicropi, and Greg Wetterau—especially Greg's delicious beer brewed with purified sewage.

I would also like to thank the friends, family, and colleagues who kindly devoted many precious hours to peer-reviewing the manuscript, including the frank feedback and meticulous editing I received from my sons, Reid and Nicholas. Many other important edits poured in from a long list of colleagues and friends, starting with Lynn Broaddus, Debbie Cervenka, Valerie Damstra, Dan Egan, Eric Ebersberger, Jennifer Gregor, Emma Holtan, Mike Kohlman, Bonnie Matuseski, Paige Penningroth, Dave Spratt, and Erik Streed. Many others reviewed the manuscript in confidence. Thanks to those salient peer reviews, this book is much stronger than it otherwise would have been. Others provided helpful feedback on the book proposal, including James Blue, Lynn Broaddus, Peter Gove, Theron O'Connor, Ry Rivard, Dave Spratt, and Brad Stone.

Purified is my third book with Island Press. Early in the book proposal stage, I reached out to David Miller, the publisher's president. His gracious and cerebral feedback was extremely helpful in transitioning this project from a bag of stories into a comprehensive narrative. I would also like to thank Erin Johnson, my editor. She provided consistent, patient, and thoughtful feedback on the manuscript, the graphics, the title, the cover, and everything else that goes into the sausage-making of publishing. Her supportive feedback was a comfort throughout.

Few had a bigger impact on the book than my colleague Valerie Damstra at the Mary Griggs Burke Center for Freshwater Innovation at Northland College. She led our team of outstanding researchers, including Emma Holtan, Paige Penningroth, Kristen Vensland, and Elsie Dickover, who helped build one of the most comprehensive water recycling archives in the United States. But that was just the beginning. Val worked closely with Jerry Lehman, the talented designer who created most of the graphics in this book. She also conducted much of the research, secured the rights and permissions, helped with fact-checking, and juggled myriad other responsibilities. In short, Val was my go-to person, and she always came through.

Also on the research side, I'd like to thank Julia Waggoner and Elizabeth Madsen-Genszler from Northland College's Dexter Library for their prompt and thoughtful responses to our numerous research requests.

I would also like to thank the rest of my Burke Center colleagues. We are a tight-knit group dedicated to wedding the disciplines of science and communication on behalf of a better-informed and ecologically engaged public. That vision came from Michael Miller, former president of Northland College, who founded the Burke Center in 2015. Now retired, he remains a good friend and enthusiastic supporter of our work. I would also like to thank Karl Solibakke and Chad Dayton, who served as consecutive presidents at Northland while I was writing this book and proved to be gracious and unwavering in their support.

Debbie Cervenka has long been a steadfast advocate for my water research and writing. A close friend and passionate conservationist, she has strong connections to many of the places highlighted in the preceding pages, including the Great Lakes, Florida, and the Southwest.

I would also like to thank the Institute for Journalism and Natural Resources for some early and instrumental support of my research on the Colorado River, as well as Buddy Huffaker and Curt Meine at the Aldo Leopold Foundation, who were particularly helpful in assisting my archival research on Leopold's historic paddle of the Colorado river delta in 1922.

Many others helped with the reporting for this book, but they are not quoted in the preceding pages. I would like to thank some of them here, including Patti Aaron, Robert Beltran, John Bonsangue, Charles Bott, Daniel Bunk, Carrie Capuco, Heather Collins, Liz Crosson, Todd Darden, Amy Dorman, Zachary Dorsey, Bill Dunivin, Robyn Felix, Daniel Fischer, Karl Flessa, Greg Fogel, Philip Friess, Terry Fulp, Sharon Green, Peter Grevatt, Ben Grumbles, Drew Hammond, Bill Hasencamp, Robert Hernandez, Warren Hogg, Dan Holloway, Perry Kaye, Sandra Kerl, Rebecca Kimitch, Michael King, Steve Krai, Paul Liu, Bronson Mack, Jennifer Marroquin, Mark Millan, Thomas Minwegen, Jamie Mitchell, Brandon Moore, David Nelms, John Nielsen-Gammon, Eric Owens, Brian Owsenek, Denise Parra, Colby Pellegrino, Gary Rasp, Scott Reinert, Leila Rice, Mia Rose Wong, Art Ruiz, Monica Sanchez, Susie Santilena, Sandy Scott-Roberts, Rupam Soni, Patricia Tennyson, Shannon Thomason, Crystal Thompson, David Todd, Rafael Villegas, and Wayne Young.

Finally, it can be difficult to be married to a journalist. No one knows that more than my wife, Meri. From my decade working as a correspondent for *Newsweek* to the last twenty years writing books, the work has regularly ended up being more difficult than it needed to be—on the entire family. Throughout it all, Meri has been unfailingly tolerant, exceptionally helpful, and exceedingly supportive. A superlative soulmate, she has consistently helped ensure that my work is the best that it can be. A beautiful writer and award-winning graphic designer, she also conceptualized the cover design for this book. Her sound instincts and patient, knowing, counsel have heavily influenced my writing for years, including this work. Most important of all, her loving support always has—and always will—leave me humbled.

Notes

1. Dead Pool

1. Chelsea Harvey, "Western 'Megadrought' Is the Worst in 1,200 Years," *Scientific American*, February 15, 2022, https://www.scientificamerican .com/article/western-megadrought-is-the-worst-in-1-200-years/.

2. Sam Hart, "Lake Mead at a Low," Reuters, August 9, 2021, https://www .reuters.com/graphics/USA-CLIMATE/DROUGHT/dwpkrgbqovm/.

3. National Park Service, "The Third Straw," accessed January 10, 2022, https://www.nps.gov/lake/learn/the-third-straw.htm.

4. Bronson Mack, spokesman, Southern Nevada Water Authority, interview with author, May 13, 2021.

5. Minimum power pool happens before dead pool, meaning that the turbines will stop turning before dead pool is actually reached. During minimum power pool, some water still slips past the dam, but there's not enough head, or pressure, to turn the turbines. Dead pool happens when reservoir levels drop so far that water no longer passes the dam.

6. Bureau of Reclamation, "Frequently Asked Questions and Answers," Lower Colorado Region, Hoover Dam, accessed January 14, 2022, https://www.usbr.gov/lc/hooverdam/faqs/powerfaq.html.

7. Tim Fitzsimons, "Historically Low Water Levels in Lake Mead Expose Intake Valve," *NBC News*, April 29, 2022, https://www.nbcnews.com /news/us-news/historically-low-water-levels-lake-mead-expose-intake -valve-rcna26674.

8. Christian Martinez, "Skeleton in Barrel Revealed by Receding Waters of Lake Mead Are of a Gunshot Victim, Police Say," *Los Angeles Times*, May 3, 2022, https://www.latimes.com/california/story/2022-05-03/remains -uncovered-by-receding-water-levels-in-lake-mead-died-from-gunshot -wound-police-say.

9. David Wilson, "Gun Found at Lake Mead Near Site Where Body Was Discovered," *Las Vegas Review-Journal*, August 18, 2022, https://www .reviewjournal.com/crime/gun-found-at-lake-mead-near-site-where-body -was-discovered-2625747/.

10. "As Lake Mead's Water Recedes, Theories Emerge about Identity of Human Remains Found in Barrel—and Possible Mob Links," unsigned article, *CBS News*, July 5, 2022, https://www.cbsnews.com/news /lake-mead-water-level-human-remains-identity-theories-mobsters/.

11. Harvey, "Western 'Megadrought' Is the Worst in 1,200 Years."

12. United States Department of Interior Bureau of Reclamation, "Colorado River Basin Water Supply and Demand Study," Executive Summary, December 2012, 9.

13. Arizona Department of Water Resources, accessed May 22, 2023, https:// www.arizonawaterfacts.com/water-your-facts.

14. Ian James, *Los Angeles Times*, "Where Colorado River No Longer Meets the Sea, a Pulse of Water Brings New Life," June 23, 2022, accessed May 22, 2023, https://www.latimes.com/california/story/2022-06-23/water -is-flowing-again-in-mexicos-dry-colorado-river-delta.

15. Camille Touton, commissioner, United States Bureau of Reclamation, accessed May 22, 2023, https://www.energy.senate.gov/services/files/6CB 52BDD-57B8-4358-BF6B-72E40F86F510.

16. Ian James, "Breakthrough Colorado River Deal Reached, Bringing Big Water Cuts for Three Years," *Los Angeles Times*, May 22, 2023, accessed, May 22, 2023, https://www.latimes.com/environment/story/2023-05-22 /seven-states-announce-colorado-river-water-deal-agreeing-on-water-cuts -for-three-years.

17. Ian James, "Colorado River Water Deal: Is it Enough to Stave Off Disaster?," *Los Angeles Times*, May 22, 2023, accessed May 22, 2023, https:// www.latimes.com/environment/story/2023-05-22/a-bold-plan-for -colorado-river-water-crisis-paying-1-billion-not-to-grow-crops.

18. "Lake Powell Needs 15 More Years of Snowpack," unsigned article,

9 News, February 9, 2023, https://www.9news.com/video/news/state/colorado-climate/lake-powell-needs-15-more-years-of-snowpack/73-52823f8e-24fc-4f80-887e-21d7a94d4c6c.

19. James Lawrence Powell, *Dead Pool: Lake Powell, Global Warming, and the Future of Water in the American West* (Berkeley: University of California Press, 2008), 22.

20. Arizona Water Banking Authority, accessed May 22, 2023, waterbank.az.gov.

21. Southern Nevada Water Authority, "Understand Laws and Ordinances," accessed May 22, 2023, https://www.snwa.com/importance-of-conservation/understand-laws-ordinances/index.html#:~:text=Replacing%20useless%20grass%20(AB356),nonfunctional%20grass%2C%20beginning%20in%202027.

22. California Water Resources Control Board, "Existing Seawater Desalination Facilities," accessed May 23, 2023, https://www.waterboards.ca.gov/water_issues/programs/ocean/desalination/docs/170105_desal_map_existing.pdf.

23. Hayley Smith, "California Coastal Commission Oks Desalination Plant in Orange County," *Los Angeles Times*, October 13, 2022, https://www.latimes.com/california/story/2022-10-13/california-coastal-commission-oks-desalination-plant-in-orange-county.

24. Florida Department of Environmental Protection, "Florida Water Resource Caution Areas," accessed May 22, 2023, https://geodata.dep.state.fl.us/datasets/f51b88085e5e4e6f892dda74eedd4789/explore?location=26.983859%2C-81.756391%2C5.75.

25. Florida Department of Environmental Protection, "One Water Florida," accessed January 6, 2022, https://floridadep.gov/southwest/sw-permitting/campaign/one-water-florida.

26. Lynn Spivey, "The Progression of Potable Reuse in Florida and the Impact of WateReuse Florida," *WateReuse Review*, accessed November 12, 2022, https://watereuse.org/2022-potable-reuse-progress-and-watereuse-florida/.

27. Hampton Roads Sanitation District, "HRSD Breaks Ground on First Full-Scale Swift Facility," July 21,2022, accessed May 22, 2023, https://www.hrsd.com/news-release-july-21-2022.

28. Robert Glennon, "Los Angeles Needs to Reclaim What We Used to Consider 'Wastewater,'" *Los Angeles Times*, March 5, 2019.

29. Stephen Smith and Samara Freemark, "Thirsty Planet, Israel: Using Technology, Engineering to Cut Reliance on Galilee," *American Public Media*, May 12, 2016, https://www.apmreports.org/story/2016/05/20/water-israel.

30. European Parliament, "Parliament Approves Increased Water Reuse," May 13, 2020, https://www.europarl.europa.eu/news/en/press-room/20200512IPR78921/parliament-approves-increased-water-reuse.

31. Rabia Chaudhry, US Environmental Protection Agency, virtual presentation at the WateReuse Association annual meeting, March 16, 2021.

2. "Gulp!"

1. Kathryn Balint, "Water from (Gulp!) Where? City Aims to Make Sewage Drinkable," *San Diego Union-Tribune*, July 6, 1997.

2. Balint, "Water from (Gulp!) Where?"

3. Tony Perry, "San Diego Stadium Expansion Foe Becomes Designated Villain," *Los Angeles Times*, February 11, 1997, https://www.latimes.com/archives/la-xpm-1997-02-11-mn-27606-story.html.

4. Balint, "Water from (Gulp!) Where?"

5. United States Navy, "Welcome to Naval Base San Diego," accessed February 13, 2022, https://cnrsw.cnic.navy.mil/Installations/NAVBASE-San-Diego/.

6. San Diego Regional Economic Development Corporation, "San Diego's Defense Cluster," March 2018, https://www.sandiegobusiness.org/sites/default/files/Defense_0.pdf.

7. Farm Bureau San Diego County, "Why Is Farming Important to San Diego County?," accessed February 11, 2022, https://www.sdfarmbureau.org/san-diego-agriculture/.

8. San Diego Regional Economic Development Corporation, "San Diego's Tourism Economy," March 2018, https://www.sandiegobusiness.org/sites/default/files/Tourism_0.pdf.

9. City of San Diego, "Population," accessed February 20, 2022, https://www.sandiego.gov/economic-development/sandiego/population.

10. City of San Diego Water Reuse Study 2005, American Assembly Workshop I—Water Reuse Goals, Opportunities and Values, October 6–7, 2004, 23.

11. City of San Diego, "Water Supply," accessed February 18, 2022, https://www.sandiego.gov/public-utilities/sustainability/water-supply.

12. San Diego County Water Authority, "Your Water," accessed February 18, 2022, https://www.sdcwa.org/your-water/.

13. City of San Diego Water Reuse Study 2005, 10.

14. City of San Diego Water Reuse Study 2005, 10.

15. City of San Diego Water Reuse Study 2005, 11.

16. The water reuse initiative was part of a complex legal settlement with the US Environmental Protection Agency. The EPA wanted San Diego to upgrade the city's aging wastewater treatment plant, which was dumping 175 million gallons of nominally treated sewage into the Pacific every day. San Diego balked at the $3 billion price tag, arguing that the sewage was being discharged into waters that were so deep, and so far out to sea, that it didn't matter. The EPA sued, and as part of the settlement, San Diego was given a rare Clean Water Act sewage-discharge waiver—as long as it invested heavily in nonpotable water recycling.

17. Kathryn Balint, "Reclaimed Water: An Asset or Just a Pipe Dream?," *San Diego Union-Tribune*, April 16, 1995.

18. Kathryn Balint, "Plan to Turn Wastewater into Drinking Water: Sewer Water Can Be Made Fit for Faucet, Opponents Cry Foul," *San Diego Union-Tribune*, December 7, 1997.

19. Matt Potter, "Can Bruce Henderson Be Detained at Guantanamo until This Is All Over? Who Will Be the Last Man Standing in Chargers Stadium Fight?," *San Diego Reader*, February 10, 2016.

20. Tony Perry, "San Diego Stadium Expansion Foe Becomes Designated Villain," *Los Angeles Times*, February 11, 1997, https://www.latimes.com/archives/la-xpm-1997-02-11-mn-27606-story.html.

21. Balint, "Plan to Turn Wastewater into Drinking Water."

22. Kathryn Balint, "Reclaimed Wastewater Viewed as Way to Quench Growing Need," *San Diego Union-Tribune*, April 3, 1994.

23. Western Consortium for Public Health, "Health Effects Study: City of San Diego Total Resource Recovery Project," 1997.

24. Peter Rowe, "San Diego's Water: Choose Your Poison," *San Diego Union-Tribune*, July 13, 1997.

25. Phaedra S. Corso, "Costs of Illness in the 1993 Waterborne *Cryptosporid-*

ium Outbreak, Milwaukee, Wisconsin," *Emerging Infectious Diseases*, April 2003, 426–31.

26. Kathryn Balint, "Plan to Turn Wastewater into Drinking Water."

27. Kathryn Balint, "In Some Neighborhoods, You'll Get the Effluent of the Affluent," *San Diego Union-Tribune*, December 7, 1997.

28. Kathryn Balint, "Plan to Use Reclaimed Water Stirs Hot Debate, Special Assembly Hearing Draws 250," *San Diego Union-Tribune*, December 9, 1997.

29. Kathryn Balint, "In Some Neighborhoods You'll Get the Effluent of the Affluent."

30. Elizabeth Royte, "A Tall, Cool Drink of … Sewage?," *New York Times Magazine*, August 8, 2008.

31. Logan Jenkins, "Tap Water via Toilets? Big Money," *San Diego Union-Tribune*, December 8, 1997.

32. Neil Morgan, "Fitting Water into Puzzle of Why S.D. Doesn't Run," *San Diego Union-Tribune*, December 14, 1997.

33. National Research Council, *Issues in Potable Reuse: The Viability of Augmenting Drinking Water Supplies with Reclaimed Water* (Washington, DC: National Research Council, 1998), 3.

34. National Research Council, *Issues in Potable Reuse*.

35. Kathryn Balint, "Sewage Solutions, Toilet-to-Tap Plan Might Be Delayed, Whole City Would Get Wastewater in 2005," *San Diego Union-Tribune*, November 12, 1998.

36. Ray Huard, "Toilet-to-Tap Water Proposal Is Placed on Hold by Golding," *San Diego Union-Tribune*, December 8, 1998.

37. Kathryn Balint and Gerry Braun, "Public Distaste Stalls 'Toilet to Tap,' Future Appears Murky for Sewage Purification Project," *San Diego Union-Tribune*, December 9, 1998.

38. Balint and Braun, "Public Distaste."

39. Balint and Braun, "Public Distaste." Unfortunately, Stevens passed away before research on the book started.

40. Balint and Braun, "Public Distaste."

41. Ray Huard, "Toilet-to-Tap Idea Gets the Big Flush, This Time for Good," *San Diego Union-Tribune*, May 19, 1999.

42. Huard, "Toilet-to-Tap Idea."

43. Suzanne Kenney, "Purifying Water: Responding to Public Opposition to the Implementation of Direct Potable Reuse in California," *UCLA Journal of Environmental Law and Policy* 37, no. 1 (2019): 119.

44. "New Survey Reveals Californians' Overwhelming Support for Recycled Water as a Long-Term Solution," *Business Wire*, March 16, 2016.

45. Rob Davis, "We Have a Winner…," *Voice of San Diego*, December 4, 2007.

46. Rob Davis, "Reuse Veto Vote," *Voice of San Diego*, December 3, 2007.

47. Davis, "We Have a Winner.…"

3. Orange County Sets the Bar

1. Mike Lee, "S.D. Looks North for Help Marketing Recycled Water, Orange County Staged Successful PR Campaign," *San Diego Union-Tribune*, September 12, 2005.

2. Orange County Water District, "A History of Orange County Water District," 2014, 11.

3. Orange County Water District, "A History," 29.

4. Frank Mickadeit, "The Danger: Barriers Keep Ocean at Bay and Freshwater Supplies Safe," *Orange County Register*, January 17, 1988.

5. "Concerned about Future Water in Orange County?," newspaper advertisement, *Los Angeles Times*, December 1, 1998.

6. Harrison Sheppard, "Some Want More Aggressive Plan to Purify Waste Water," *Los Angeles Times*, December 9, 1998.

7. US Environmental Protection Agency, "Technical Fact Sheet—N-Nitrosodimethylamine (NDMA)," EPA 505-F-17-005, November 2017, https://www.epa.gov/sites/default/files/2017-10/documents/ndma_fact_sheet_update_9-15-17_508.pdf.

8. Pat Brennan and Gary Robbins, "Carcinogen Found in O.C. Wells, Agency Says the Chemical's Presence Is Minimal and the Water Supply Is Safe," *Orange County Register*, May 31, 2000.

9. Pat Brennan, "Contaminant Shuts Water Wells, Officials Say There Is Not Threat to Public Health and that Nine Closures Are Precautionary," *Orange County Register*, January 30, 2002.

10. Grace Camacho, "Wells Remain Shut Down until Treatment Decision

Made, A Cancer-Causing Compound Is Found in Drinking Water," *Orange County Register*, February 7, 2002.

11. Jason Dadakis, "Technical Memorandum," Orange County Water District, August 7, 2013. Memorandum was sent to the National Water Resources Research Institute's Independent Advisory Panel monitoring the Groundwater Replenishment District program.

12. Missouri Department of Health and Senior Services et al., "The Sewershed Surveillance Project," accessed March 29, 2022, https://storymaps .arcgis.com/stories/f7f5492486114da6b5d6fdc07f81aacf.

13. Brian Bernados, technical specialist, California State Water Resources Control Board, presentation, California WateReuse Conference, San Francisco, September 13, 2022.

14. Bernados, presentation.

15. Orange County Water District, "Final Expansion," accessed May 15, 2023, https://www.ocwd.com/gwrs/final-expansion/.

4. San Diego Bounces Back

1. National Oceanic and Atmospheric Administration, "What Is a Thermocline?," accessed January 19, 2023, https://oceanservice.noaa.gov/facts /thermocline.html. See also, City of San Diego, "Fact Sheet, Point Loma Wastewater Treatment Plant Secondary Equivalent," accessed May 23, 2023, https://www.sandiego.gov/sites/default/files/legacy/water/pdf/pure water/2014/ptlomafactsheet.pdf.

2. Robert Simmons, "Our Region Needs Repurified Water," *San Diego Union-Tribune*, September 11, 2001.

3. City of San Diego, "Water Reuse Study 2005: Water Reuse Goals, Opportunities and Values," American Assembly Workshop I, October 6–7, 2004, 13, https://www.sandiego.gov/sites/default/files/legacy/water /pdf/purewater/aa1wp.pdf.

4. Jose Luis Jiménez, "Few Thirst for Recycled Tap Water, Survey Says," *San Diego Union-Tribune*, August 16, 2004.

5. Mike Lee, "'Repurified' Wastewater Backed for Home Use, Citizens Panel Forwards Proposal to S.D. Council," *San Diego Union-Tribune*, July 15, 2005.

6. Marsi Steirer, "City Studies Recycled Water Opportunities," letter to the

editor, *San Diego Union-Tribune*, August 27, 2004, https://www.sandiego
.gov/sites/default/files/legacy/water/pdf/purewater/040827.pdf.

7. Mike Lee, "S.D. Looks North for Help Marketing Recycled Water, Orange County Staged Successful PR Campaign," *San Diego Union-Tribune*, September 12, 2005.

8. City of San Diego, "Water Reuse Study, Final Draft Report," March 2006, 4-2–4-6, 7-39, https://www.sandiego.gov/sites/default/files/legacy /water/purewater/pdf/ir06toc.pdf.

9. Rob Davis, "Not Too Freaked Out about Recycled Sewage," *Voice of San Diego*, November 18, 2008, https://voiceofsandiego.org/2008/11/18/not -too-freaked-out-about-recycled-sewage/. See also Mike Lee and Jennifer Vigil, "Council Beats Veto on Water Recycling," *San Diego Union-Tribune*, December 4, 2007.

10. "Rate Hike Caveats, 'Toilet to Tap,' Sewage Plant Overhaul Must Go," unsigned editorial, *San Diego Union-Tribune*, November 22, 2006.

11. "The Yuck Factor: Get Over It," unsigned editorial, *San Diego Union-Tribune*, January 23, 2011, https://www.sandiegouniontribune.com/opin ion/editorials/sdut-the-yuck-factor-get-over-it-2011jan23-story.html.

12. Mike Lee, "Backers of Water Recycling See a Rising Tide of Support," *San Diego Union-Tribune*, May 16, 2011.

13. National Research Council, *Issues in Potable Reuse: The Viability of Augmenting Drinking Water Supplies with Reclaimed Water* (Washington, DC: National Research Council, 1998), 3.

14. National Academy of Sciences, Engineering, Medicine, "Reuse of Municipal Wastewater Has Significant Potential to Augment Future U.S. Drinking Water Supplies," January 12, 2012.

15. "Okun Remembered for Pioneering Work in Water Engineering (Spring, 2008)," unsigned article, *Carolina Public Health Magazine*, December 11, 2007, https://sph.unc.edu/cphm/carolina-public-health-magazine-build ing-and-inspiring-leaders-spring-2008/okun-remembered-for-pioneering -work-in-water-engineering-spring-2008/.

16. Mike Lee, "Recycled Water Getting Another Look," *San Diego Union-Tribune*, May 23, 2012.

17. Chelsea Harvey, "Western 'Megadrought' Is the Worst in 1,200 Years," *Scientific American*, February 15, 2022, https://www.scientificamerican .com/article/western-megadrought-is-the-worst-in-1-200-years/.

18. Ian James, "Western Megadrought Is Worst in 1,200 Years, Intensified by Climate Change, Study Finds," *Los Angeles Times*, February 14, 2022, https://www.latimes.com/environment/story/2022-02-14/western-mega drought-driest-in-1200-years.

19. Deborah Sullivan Brennan, "Tide Turns in Favor of Recycling Wastewater," *San Diego Union-Tribune*, April 29, 2013.

20. Liam Dillon, "San Diegans Are Already Drinking Pee," *Voice of San Diego*, June 23, 2014, https://voiceofsandiego.org/2014/06/23/san-diegans-are -already-drinking-pee/.

21. David Garrick, "Council Approves Sewage Recycle Program," *San Diego Union-Tribune*, November 19, 2014.

22. Amy Dorman, project delivery manager, Pure Water San Diego, City of San Diego Public Utilities Department, interview with author, January 4, 2023.

23. Deborah Sullivan Brennan, "Council OK Could Open Floodgates for Water Purification," *San Diego Union-Tribune*, November 16, 2014.

24. Ry Rivard, "Contractors See Pure Water Case as a Test for Big Projects Across Region," *Voice of San Diego*, July 1, 2019.

25. Dorman, interview.

5. Future Water in Virginia

1. Interview with David Nelms, retired hydrologist, United States Geological Survey, May 19, 2023.

2. Jack Eggleston and Jason Pope, "Land Subsidence and Relative Sea-Level Rise in the Southern Chesapeake Bay Region," United States Geological Survey, Circular 1392, 2013, 1.

3. Eggleston and Pope, "Land Subsidence," 15, 20.

4. Dan Holloway, hydrogeologist, Hampton Roads Sanitation District, email correspondence with author, September 22, 2022. See also National Oceanic and Atmospheric Administration, "Relative Sea Level Trend 8638610 Sewells Point, Virginia," accessed June 25, 2022, https://tides andcurrents.noaa.gov/sltrends/sltrends_station.shtml?id=8638610; and C. Todd Lopez, "DOD, Navy Confront Climate Change Challenges in Southern Virginia," *Department of Defense News*, July 21, 2021, https:// www.defense.gov/News/News-Stories/Article/Article/2703096/dod-navy -confront-climate-change-challenges-in-southern-virginia/.

5. National Oceanic and Atmospheric Administration, "Hampton Roads' Sea Level Rise Adaptation Advances on Multiple Fronts," accessed June 16, 2022, https://coast.noaa.gov/states/stories/sea-level-rise-adaptation-advances-on-multiple-fronts.html.

6. Forbes Tompkins and Christina Deconcini, "Sea-Level Rise and Its Impact on Virginia," World Resources Institute Fact Sheet, June 2014, 2, https://research.fit.edu/media/site-specific/researchfitedu/coast-climate-adaptation-library/united-states/east-coast/virginia/Tompkins--DeConcini.--2014.--SLR--its-Impact-on-Virginia.pdf.

7. Modeling study produced by Aquaveo, LLC for Hampton Roads Sanitation District and provided by Dan Holloway, hydrogeologist, Hampton Roads Sanitation District, "Aquifer Replenishment System, VAHydro-GW Phase 2B," May 15, 2016. Graphic on the study's page 65 shows potential rise of three to six inches. Other graphics suggest even more potential rebound of the land surface.

8. Modeling study, Aquaveo, "Aquifer Replenishment System."

9. Jamie Mitchell, chief of technical services, Hampton Roads Sanitation District, email correspondence with author, September 15, 2022. At full build-out, SWIFT could end up reducing nutrients by as much as 85 percent by 2040.

10. Hampton Roads Sanitation District, "HRSD Breaks Ground on First Full-Scale Swift Facility," July 21, 2022, accessed May 23, 2023, https://www.hrsd.com/news-release-july-21-2022.

11. Scott Kudlas, director of the Office of Water Supply, Virginia Department of Environmental Quality, email correspondence with author, September 9, 2022, and May 25, 2023.

12. Marcia Degen, environmental technical services manager, Virginia Department of Health, virtual presentation, WateReuse Symposium, March 2021.

13. Ryder Bunce, SWIFT technical services engineer, Virginia Department of Health, virtual presentation WateReuse Symposium, March 2021.

14. Kudlas, email correspondence.

15. Brian Owsenek, deputy executive director, Upper Occoquan Service Authority, interview with author, June 14, 2022. Sources at Fairfax Water said that the percentage can be even higher.

16. Owsenek, interview.

17. "Pollution Is Called Dangerous," unsigned article, *Washington Post and Times Herald*, February 27, 1956.

18. "Two Officials Rap Fairfax Plan to Dump Treated Sewage in Area Water Services," unsigned article, *Washington Post and Times Herald*, November 27, 1959.

19. Noman M. Cole Jr., "Chronology of Major Events Associated with the Development of the State Water Control Board's Occoquan Policy and Development of the Upper Occoquan Sewage Authority Project," February 1, 1979. Provided by Brian Owsenek, deputy executive director, Upper Occoquan Service Authority.

20. "Moratorium Is Declared on Occoquan Sewage Plants," unsigned article, *Washington Post and Times Herald*, October 26, 1963.

21. George Lardner Jr., "Virginia Officials Undecided about Next Move after Moratorium on Occoquan Sewage Effluent," *Washington Post and Times Herald*, November 3, 1963.

22. Kenneth Bredemeier, "Work to Improve Reservoir Urged," *Washington Post and Times Herald*, April 29, 1970.

23. Kenneth Bredemeier, "Va. Water Board Proposes Occoquan Area Sewage Plant," *Washington Post and Times Herald*, March 27, 1970.

24. Associated Press, "New Standards for Wastershed Discussed," *The Bee* (Danville, Virginia) April 1, 1971.

25. Owsenek, interview.

26. Owsenek, interview.

27. US Environmental Protection Agency, "Overview of Drinking Water Technologies," accessed April 28, 2023, https://www.epa.gov/sdwa/over view-drinking-water-treatment-technologies.

28. Melissa Eddy, "The World Economy Is Slowing More Than Expected, a New Forecast Shows," *New York Times*, September 26, 2022, https://www.nytimes.com/2022/09/26/business/global-economy-oecd-forecast.html?smid=em-share.

29. Felicia Marcus, former chair of the California State Water Control Board, virtual presentation, WateReuse Symposium, March 2021.

6. Running Dry (Almost) in Texas

1. Bobby Weaver, "The Well That Launched the Permian," *Permian Basin*

Petroleum Association Magazine, October 12, 2017, https://pboilandgas magazine.com/the-well-that-launched-the-permian/; Thomas Smith, "Midland, Texas: Gateway to the Permian Basin," *GEO ExPro*, December 2, 2015, https://www.geoexpro.com/articles/2015/02/midland-texas -gateway-to-the-permian-basin.

2. Texas Water Development Board, "Drought in Texas: A Comparison of the 1950–1957 and 2010–2015 Droughts," February 2022, E1, https:// www.twdb.texas.gov/publications/reports/other_reports/doc/Drought -in-Texas-Comparison-1950s-2010s.pdf.

3. Kate Galbraith, "A City Built on Oil Discovers How Precious Its Water Can Be," *New York Times*, April 21, 2011, https://www.nytimes.com /2011/04/22/us/22ttwater.html.

4. John Womack, systems operations manager, Colorado River Municipal Water District, interview with author, May 2, 2022.

5. Ron Alton, manager, Big Spring State Park, interview with author, May 2, 2022. See also *Big Spring State Park Scenic Mountain Trail Guide*, Texas Parks and Wildlife Department, October 1992, 13–15.

6. Texas Water Development Board, "Major and Historical Springs of Texas," Report 189, March 1975, 5, https://www.twdb.texas.gov/publica tions/reports/numbered_reports/doc/R189/R189.pdf.

7. Texas Water Development Board, "Major and Historical Springs of Texas," 9.

8. Colorado River Municipal Water District, "Lake J.B. Thomas," accessed August 19, 2022, https://www.crmwd.org/water-sources/surface-water/.

9. Andreia Medlin, "Water Woes Alleviated Somewhat," *Big Spring Herald*, September 26, 2013.

10. According to its home page, the Colorado River Municipal Water District (CRMWD) produces approximately nineteen billion gallons of water per year. Nineteen billion divided by 365 days comes to an average of fifty-two million gallons per day. CRMWD homepage captured May 18, 2023, https://www.crmwd.org/about/.

11. David A. Johnson, "West Texas Communities Deal with Water Crisis," *Lubbock Avalanche-Journal*, September 26, 2011, https://www.lubbock online.com/story/business/agricultural/2011/09/27/west-texas-commun ities-deal-water-crisis/15210040007/.

12. Womack, interview.

13. Shannon Thomason, former mayor, Big Spring, Texas, interview with author, May 2, 2022.

14. John Ingle, "No Water to Waste, Big Spring Getting Used to Treated Effluent," *San Angelo Standard*, March 13, 2014.

15. Ingle, "No Water to Waste."

16. One reservoir, Lake Thomas, does not suffer from high salinity rates.

17. Brian Bernados, technical specialist, California Division of Drinking Water, presentation, WateReuse California annual conference, San Francisco, September 13, 2022.

18. Bernados, presentation.

19. Todd Darden, city manager, Big Spring, Texas, interview with author, May 2, 2022.

20. Womack, interview.

21. Russell Gibson et al., "The Emergency $130-Million Ward County Water Supply Project," *Pipelines 2013* (June 2013): 440–50, https://doi.org/10.1061/9780784413012.041.

22. Betsy Blaney, "Wichita Falls Awaits OK for Wastewater Reuse," *Austin American-Statesman*, April 13, 2014. See also John Ingle, "Chamber Meeting Addresses Drought, WF Drought Being Called Unprecedented," *Wichita Falls Times-Record-News*, April 18, 2014.

23. Letter from Texas Commission on Environmental Quality to City of Wichita Falls, June 27, 2014, 2.

24. US Environmental Protection Agency, "2017 Potable Reuse Compendium," figure A.6-5, https://www.epa.gov/sites/default/files/2018-01/documents/potablereusecompendium_3.pdf.

25. Audrey White, "Water-Reuse Ideas Go Forward Despite 'Toilet to Tap' Concerns," *New York Times*, February 7, 2013, https://www.nytimes.com/2013/02/08/us/potable-water-reuse-ideas-go-forward-in-texas-despite-concerns.html.

26. Aaron Galloway, "Jimmy Fallon Cracks Joke about Wichita Falls during Tonight Show Monologue [VIDEO]," *92.9 NIN*, May 8, 2014, https://929nin.com/jimmy-fallon-makes-joke-about-wichita-falls/.

27. Deanna Watson, "TRN's Ingle Wins National Award for Water Coverage," *Wichita Falls Times-Record-News*, January 27, 2017, https://www

.timesrecordnews.com/story/news/local/2017/01/27/trns-ingle-wins
-national-award-water-coverage/97139038/.

28. John Ingle, "Pipeline Begins Final Days of Testing, Cypress Plant Could
Be Online in April," *Wichita Falls Times-Record-News*, January 28, 2014.
This is one of several stories from the series.

29. Steve Campbell, "Dry Wichita Falls to Try Drinking 'Potty Water,'" *Fort
Worth Star-Telegram*, March 14, 2014.

30. Jason Feller, "Water Reuse Project Making Wichita Falls Famous," *Wichita Falls Times-Record-News*, May 8, 2014.

31. US Environmental Protection Agency, "2017 Potable Reuse Compendium," A.7-7, https://www.epa.gov/sites/default/files/2018-01/documents
/potablereusecompendium_3.pdf.

7. El Paso's Quiet Leadership

1. National Park Service, "Chihuahuan Desert Ecoregion," accessed August
12, 2022, https://www.nps.gov/im/chdn/ecoregion.htm.

2. "Engineers Issue Report on Well Recharging Here," unsigned article, *El
Paso Times*, July 21, 1952.

3. "Must Plan Ahead for Water," unsigned article, *El Paso Times*, March 20,
1964.

4. "EP Water Situation Discussed," unsigned article, *El Paso Times*, May
23, 1969. See also Gregory Jones, "Will the Well Run Dry for Future El
Paso Residents?," *El Paso Times*, June 6, 1976; and Susan Ihne, "El Paso's
Choices: Find Water or Ration," *El Paso Times*, September 7, 1980.

5. Jones, "Will the Well Run Dry?"

6. Paul Sweeney, "Plant to Recycle Wastewater," *El Paso Times*, May 11,
1978. See also "Northeast Water Recycling Plant Set to Begin Operating
This Week," unsigned article, *El Paso Times*, June 2, 1985.

7. US Environmental Protection Agency, "Overview of Drinking Water
Treatment Technologies," accessed August 17, 2022, https://www.epa.gov
/sdwa/overview-drinking-water-treatment-technologies.

8. Lisa Chamberlain, "Two Cities and Four Bridges Where Commerce
Flows," *New York Times*, March 28, 2007, https://www.nytimes.com
/2007/03/28/realestate/commercial/28juarez.html.

9. Denise Bezick, "Board OKs Water Limits for El Paso," *El Paso Times*,
November 29, 1990.

10. "Rio Grande Next Best Hope," unsigned article, *El Paso Times*, April 27, 1996.

11. "Rio Grande Next Best Hope."

12. Dan Williams, "El Paso Weighs Options as Bolson Drains," *El Paso Times*, April 27, 1996. See also *El Paso Times*, sidebar to "Rio Grande Next Best Hope," April 27, 1996.

13. Sharon Simonson, "Industries, Agriculture Already Worry about Water," *El Paso Times*, March 21, 1999. See also Vic Kolenc, "City Shuns Water-Guzzling Industries," *El Paso Times*, March 26, 2000.

14. Marty Schladen, "Quenching Our Future, Part I: Protecting the Rio Grande's Basin's Dwindling Water," *El Paso Times*, November 16, 2014, https://www.elpasotimes.com/story/archives/2014/11/15/protecting-our -dwindling-water/74070200/.

15. Ed Archuleta, "Desalination Plant Ensures El Paso's Water Needs, Groundbreaking Is Wednesday for Joint Project with Fort Bliss," *El Paso Times*, August 28, 2005.

16. Art Ruiz, superintendent, El Paso Water's Kay Baily Hutchison Desalination Plant, interview with author, April 4, 2022.

17. Ann Espinola, "How One Utility Won Public Support for Potable Reuse," *American Water Works Association Articles*, January 21, 2016, https://www .awwa.org/AWWA-Articles/how-one-utility-won-public-support-for -potable-reuse, accessed September 28, 2022.

18. US Environmental Protection Agency, "PFAS Explained," Water Topics, Per- and Polyfluoroalkyl Substances (PFAS), accessed August 4, 2022, https://www.epa.gov/pfas/pfas-explained.

19. Texas Water Development Board, "Drought in Texas: A Comparison of the 1950-1957 and 2010-2015 Droughts," February 2022, E1-E3, accessed May 18, 2023, https://www.twdb.texas.gov/publications/reports /other_reports/doc/Drought-in-Texas-Comparison-1950s-2010s.pdf.

20. Texas Water Development Board, "Drought in Texas."

21. Michael Barnes, "The Times It Never Rained: Three Devastating Historic Texas Droughts," *Austin American-Statesman*, July 18, 2022.

22. Pat H. Holland, "Diversions from Red River to Lake Dallas, Texas; and Related Channel Losses," US Department of Interior, Geological Survey, February and March 1954.

23. Julian Aguilar, "Texas Climatologists Warn of Potential for Prolonged Drought," *KERA News*, March 30, 2022, https://www.keranews.org /texas-news/2022-03-30/state-climatologists-warn-of-potential-for-pro longed-drought.

24. Dave Fehling, "Dallas Wastewater Keeps Trinity Flowing, Houston Drinking," *NPR*, December 21, 2011, https://stateimpact.npr.org/texas /2011/12/21/dallas-wastewater-keeps-trinity-flowing-houston-drinking/.

25. Dallas Water Utilities, "2014 Dallas Long Range Water Supply Plan to 2070 and Beyond," December 2015, 15-16, https://dallascityhall.com /departments/waterutilities/DCH%20Documents/2014_LRWSP_Final _Report_all_11302015.pdf.

26. Texas Water Development Board, "History of Water Reuse in Texas," February 2011, 4, https://www.twdb.texas.gov/innovativewater/reuse /projects/reuseadvance/doc/component_a_final.pdf.

27. Texas Water Development Board, "Water Supply Planning in Texas," accessed May 18, 2023, https://www.twdb.texas.gov/waterplanning /index.asp.

28. Texas Water Development Board, "Regional Water Planning Groups," accessed August 25, 2022, https://www.twdb.texas.gov/waterplanning /rwp/regions/index.asp.

8. Hot Tempers in Tampa

1. "Bigger Is Better? 1970s Documentary on Growing Pains in Tampa Bay," *60 Minutes*, produced by Norman Gorin, YouTube video, 17:13, June 26, 2018, https://www.youtube.com/watch?v=PpW6AG31yWY.

2. Macrotrends, "Tampa Metro Area Population 1950–2023," accessed May 5, 2023, https://www.macrotrends.net/cities/23160/tampa/popu lation.

3. Tampa Economic Development Council, "Net Migration Pattern by State," accessed May 5, 2023, https://tampabayedc.com/news/how-many -people-moved-to-florida-this-past-year/.

4. Southwest Florida Water Management District, "Recycled Water," accessed December 18, 2022, https://www.swfwmd.state.fl.us/projects /recycled-water.

5. Natural Resources Defense Council, "Climate Change, Water, and Risk: Current Water Demands Are Not Sustainable," July 2010, accessed May 5, 2023, https://www.nrdc.org/sites/default/files/WaterRisk.pdf.

6. Bluefield Research, "U.S. Municipal Wastewater and Reuse: Market Trends, Opportunities, and Forecasts, 2015–2025," July 2015, 10.

7. Bluefield Research, "U.S. Municipal Wastewater and Reuse."

8. Brian Armstrong, executive director, Southwest Florida Water Management District, interview with author, November 22, 2022.

9. Armstrong, interview.

10. Florida Climate Center, "Drought," May 11, 2023, https://climatecenter .fsu.edu/topics/drought.

11. Florida Department of Health, "Florida's Water-What You Should Know," June 15, 2004, accessed December 16, 2022, https://www.floridahealth .gov/environmental-health/private-well-testing/_documents/Drinking Water-IsYoursSafe.pdf.

12. J. D. Callaway, "Treated Effluent May Be Recycled into Tap Water," *Tampa Tribune*, May 24, 1986.

13. Armstrong, interview.

14. Donna M. Schiffer, "Hydrology of Central Florida Lakes: A Primer," US Geological Survey, Circular 1137, 1998, 21, accessed December 17, 2022, https://fl.water.usgs.gov/PDF_files/c1137_schiffer.pdf.

15. Honey Rand, *Water Wars: A Story of People, Politics and Power* (Bloomington, IN: Xlibris Corp., 2003).

16. Warren Hogg, chief science officer, Tampa Bay Water, interview with author, November 22, 2022.

17. Weather U.S., "Climate and Monthly Weather Forecast Tampa, FL," 2023, https://www.weather-us.com/en/florida-usa/tampa-climate.

18. Armstrong, interview; "A Plan to End the Water Wars at Last," unsigned editorial, *Tampa Tribune*, December 22, 1996.

19. The predecessor to Tampa Bay Water was called the West Coast Regional Water Supply Authority.

20. "A Plan to End the Water Wars at Last." See also Francis Gilpin, "Tampa Officials Leery of Truce in Water Wars," *Tampa Tribune*, February 25, 1998.

21. "Public Meeting on the Tampa Water Resource Recovery Project," public notice, *Tampa Tribune*, June 14, 1997.

22. "Drink Sewer Water?," paid political advertisement, *Tampa Bay Times*, July 29, 1998.

23. Jean Heller, "Recycled Water? Only in a Pinch," *Tampa Bay Times*, March 14, 1998.

24. National Academy of Sciences, Engineering, and Medicine, "Reuse of Municipal Wastewater Has Significant Potential to Augment Future US Drinking Water Supplies," January 13, 2012, 4.

25. Anne Bartlett, "Recycling Water Gaining Support," *Tampa Tribune*, August 1, 1988.

26. Richard Danielson, "Tampa to Study 'Toilet-to-Tap' Scope," *Tampa Bay Times*, June 3, 2016.

27. Danielson, "Tampa Considers Toilet-to-Tap Plan."

28. John Romano, "Water Deal May Open Spigot of Animosity," *Tampa Bay Times*, October 11, 2018.

29. Craig Pittman, "Water Decision Punted, Yet Again: Tampa Bay Water Board Votes to Put Off Making Call on 'Toilet to Tap' for a Year," *Tampa Bay Times*, April 16, 2019.

30. Charlie Frago, "On the Brink of a New Water War? Tempers Flare on a Tampa Bay Board over Tampa's Proposal for 'Toilet-to-Tap,'" *Tampa Bay Times*, August 20, 2019.

31. Sallie Parks et al., "Regional Water Plan Works," *Tampa Bay Times*, April 11, 2019.

32. Lynn Spivey, former chair of the Florida Potable Reuse Commission, interview with author, October 18, 2022.

33. PURE stood for Purify Usable Resources for the Environment. "Tampa Bay Water Monitors Tampa's PURE Project," unsigned article, *Tampa Bay Water Blog*, February 15, 2022, https://www.tampabaywater.org/blog /tampa-bay-water-monitors-tampas-pure-project/.

34. Charlie Frago, "Tampa Renews Push for Reusing Wastewater," *Tampa Bay Times*, January 12, 2021, https://www.tampabay.com/news/tampa /2021/01/12/tampa-renews-push-for-reusing-sewage-for-water-needs/.

35. Juturna Consulting, LLC, et al., "Analysis of Alternatives to Reduce Non-Beneficial Treated Wastewater Discharge, Improve Supply Reliabil- ity, and Improve Minimum Flows in the Lower Hillsborough River," July 5, 2021, 69.

36. Charlie Frago, "Tampa Wastewater Reuse Project under Fire Again," *Tampa Bay Times*, September 13, 2022, https://www.tampabay.com/news /tampa/2022/09/13/tampa-wastewater-reuse-project-under-fire-again/.

37. Frago, "Tampa Wastewater Reuse Project under Fire Again."

38. Answers to the questions were finally delivered in January 2023.

39. Charlie Frago, "Tampa's Plan for Its Wastewater in Peril after City Council Vote," *Tampa Bay Times*, September 15, 2022, https://www.tampabay .com/news/tampa/2022/09/15/tampas-plan-its-wastewater-peril-after -city-council-vote/.

40. Evan Axelbank, "Tampa City Council Throws Cold Water on Sewage Project," *FOX 13 News*, September 15, 2022, https://www.fox13news .com/news/tampa-city-council-throws-cold-water-on-sewage-project.

41. William March, "Young Republicans Target Tampa Mayor Jane Castor with Dirty Water Ad," *Tampa Bay Times*, December 16, 2022, https:// www.tampabay.com/news/tampa/2022/12/16/young-republicans-target -tampa-mayor-jane-castor-with-dirty-water-ad/.

42. At one time Castor was a Republican, but she switched her party affiliation before running for mayor.

43. March, "Young Republicans Target Tampa Mayor."

44. Hogg, interview.

45. Armstrong, interview.

9. Going Beyond Purple Pipe in Florida

1. The Florida Legislature Office of Economic and Demographic Research, *Annual Assessments of Florida's Conservation Lands, Water Resources (Quantity and Quality), Stormwater and Wastewater Facilities, and Flooding Resiliency,* January 18, 2023, 9, http://edr.state.fl.us/content/presentations /water%20resources-conservation%20lands/LandandWaterAnnualAssess ment_2023Presentation.pdf.

2. Florida Department of Environmental Protection, "Florida Water Resource Caution Areas (WRCA)," accessed December 1, 2022, https:// geodata.dep.state.fl.us/datasets/f51b88085e5e4e6f892dda74eedd4789 /explore?location=26.934951%2C-81.544133%2C5.49.

3. "Water Resource Caution Area Definition," unsigned article, *Law Insider*, https://www.lawinsider.com/dictionary/water-resource-caution-area.

4. Smithsonian, "North Atlantic Right Whale Territory," accessed December 21, 2022, https://ocean.si.edu/ocean-life/marine-mammals/north-atlantic -right-whale-territory.

5. National Oceanic and Atmospheric Administration, "North Atlantic

Right Whale," accessed December 18, 2022, https://www.fisheries.noaa .gov/species/north-atlantic-right-whale.

6. National Oceanic and Atmospheric Administration, "North Atlantic Right Whale."

7. Bluefield Research, "U.S. Municipal Wastewater and Reuse: Market Trends, Opportunities, and Forecasts, 2015–2025," July 2015, 10, https://www.bluefieldresearch.com/research/focus-report-us-municipal -wastewater-reuse-market-trends-opportunities-forecasts-2015-2025/.

8. Lynn Spivey, "The Progression of Potable Reuse in Florida and the Impact of WateReuse Florida," *WateReuse Review*, November 1, 2022, https:// watereuse.org/2022-potable-reuse-progress-and-watereuse-florida/.

9. Southwest Florida Water Management District, "2020 Potable Water Reuse Survey Analysis Report," accessed December 15, 2022, https:// www.swfwmd.state.fl.us/sites/default/files/2020%20Potable%20Water% 20Reuse%20Survey%20Analysis%20Report_0.pdf. This poll showed that 58 percent supported indirect potable reuse and 61 percent supported direct potable reuse.

10. Hillsborough County, "Aquifer Recharge Projects," accessed December 3, 2022, https://www.hillsboroughcounty.org/en/government/county-pro jects/highlighted-cip-projects/aquifer-recharge-projects.

11. City of Altamonte Springs, pureALTA, accessed December 17, 2022, https://www.altamonte.org/DocumentCenter/View/5245/pureALTA -Info-Sheet.

12. Bart Weiss, "Creating a World of New Options—and Water to Sustain It," *Tampa Tribune*, May 15, 2015.

10. LA Goes All In

1. Eric Garcetti, former mayor of the city of Los Angeles, interview with the author, November 28, 2022.

2. Hubertus Cox, division manager, Los Angeles Sanitation and Environment, interview with author, March 13, 2020.

3. Los Angeles Department of Water and Power, "Mayor Garcetti: Los Angeles Will Recycle 100% of City's Wastewater by 2035," February 21, 2019, https://www.ladwpnews.com/mayor-garecetti-los-angeles-will -recycle-100-of-citys-wastewater-by-2035/.

4. Bettina Boxall, "L.A.'s Ambitious Goal: Recycle All of the City's Sewage

into Drinkable Water," *Los Angeles Times*, February 22, 2019, https://
www.latimes.com/local/lanow/la-me-water-recycling-los-angeles-20190
222-story.html.

5. Tracy Quinn, "City of Los Angeles Announces Bold Recycled Water
   Plan," Natural Resources Defense Council, February 21, 2019, https://
   www.nrdc.org/experts/tracy-quinn/city-angeles-announces-bold-recycled
   -water-plan.

6. Monte Morin, "Turning Sewage into Drinking Water Gains Appeal as
   Drought Lingers," *Los Angeles Times*, May 24, 2015, https://www.latimes
   .com/local/california/la-me-toilet-to-tap-20150525-story.html.

7. Quinn, "City of Los Angeles Announces Bold Recycled Water Plan."

8. Nicholas Pinhey, "California's First Water Reuse Filtration Plant Was in
   Los Angeles 1930," California Water Environment Association, accessed
   January 20, 2023, https://www.cwea.org/news/californias-first-water
   -reuse-filtration-plant-in-los-angeles-1930/.

9. Pinhey, "California's First Water Reuse Filtration Plant was in Los Angeles
   1930."

10. "Basin Water Spreading Raises District Levels," unsigned article, *Los Ange-
    les Times*, October 10, 1954. See also "Move Taken to Raise L.A. Water
    Supply," *Los Angeles Times*, November 6, 1955.

11. Bert Mann, "Reclamation Plant Hailed," *Los Angeles Times*, June 28,
    1964.

12. "$1½ Million Water Reclamation Plant Proposed for Valley," unsigned
    article, *Daily News-Post* (Monrovia, CA), November 26, 1958.

13. Department of Water Resources, "California's State Water Project," *Cali-
    fornia Agencies*. Paper 214, 1997, http://digitalcommons.law.ggu.edu
    /caldocs_agencies/214.

14. Mann, "Reclamation Plant Hailed."

15. Los Angeles County Sanitation District, "Whittier Narrows Water Recla-
    mation Plant," accessed January 28, 2023, https://www.lacsd.org/services
    /wastewater-sewage/facilities/whittier-narrows-water-reclamation-plant.

16. California Ag Water Stewardship Initiative, "Use of Municipal Recycled
    Water," accessed January 17, 2023, http://agwaterstewards.org/practices
    /use_of_municipal_recycled_water/. See also WateReuse Research Foun-
    dation, "Recycled Water: How Safe Is It?," June 3, 2011, https://awpw

.assembly.ca.gov/sites/awpw.assembly.ca.gov/files/hearings/Recycled%20
Water%20-%20How%20Safe%20Is%20It.pdf.

17. Earle Hartling, water recycling coordinator, Los Angeles County Sanitation Districts, interview with author, December 1, 2022.

18. William Kahrl, "Water, Water: Everywhere a Shortage, Yes, but Who Should Conserve?," *Los Angeles Times*, April 29, 1990, https://www
.latimes.com/archives/la-xpm-1990-04-29-op-345-story.html.

19. Citizens for Clean Water, "Do You Want Reclaimed Sewer Water In Your Underground Drinking Water?," *San Gabriel Valley Tribune*, November 3, 1993.

20. Berkley Hudson, "Mixed Reviews for Water Reclamation Plan: Miller Brewery and Other Opponents of the Project Say It Could Pose Health Risks. Environmentalists and Water Agencies Embrace It as a Way to Help 'Drought-Proof' the San Gabriel Valley," *Los Angeles Times*, December 12, 1993, https://www.latimes.com/archives/la-xpm-1993-12-12
-ga-1066-story.html.

21. Tally Goldstein, "A Brouhaha on Tap," *Los Angeles Times*, July 28, 1994, https://www.latimes.com/archives/la-xpm-1994-07-28-ga-20737-story
.html.

22. Forest Tennant, "We Don't Want Sewer Water in Our Drinking Water!" *San Gabriel Valley News*, advertisement, November 17, 1993.

23. Frank Clifford, "Storm Brews Over Prospect of Recycled Water in Beer: Miller Co. in Irwindale is Suing to Halt $25-Million Project. District Dismisses Concerns," *Los Angeles Times*, September 14, 1994, https://
www.latimes.com/archives/la-xpm-1994-09-14-mn-38525-story.html.

24. Clifford, "Storm Brews."

25. Clifford, "Storm Brews."

26. Miller Brewing Company, "Setting the Record Straight," advertisement, *Los Angeles Times*, September 8, 1994.

27. Hugh Dellios, "Brewer Fights to 'Stand Clear' of Recycled Water," *Chicago Tribune*, November 23, 1994, https://www.chicagotribune.com/news
/ct-xpm-1994-11-23-9411230204-story.html.

28. Richard Winton, "Brewery Victory May Kill Water Reclamation Plan," *Los Angeles Times*, December 13, 1994.

29. Abigail Goldman, "Smaller Water Reclamation Plan Appeases Miller Beer,"

Los Angeles Times, February 9, 1996, https://www.latimes.com/archives/la-xpm-1996-02-09-me-34069-story.html.

30. Myron Levin, "Treated Effluent May Help Fill Drinking-Water Wells: The East Valley Project Would Extend to Irrigation and Industry and Could Eventually Meet 7% of L.A.'s Water Needs," *Los Angeles Times*, September 29, 1990, https://www.latimes.com/archives/la-xpm-1990-09-29-me-882-story.html.

31. Myron Levin, "Treated-Water Proposal Gets Mixed Response," *Los Angeles Times*, October 4, 1990, https://www.latimes.com/archives/la-xpm-1990-10-04-me-2026-story.html.

32. Tim May, "Water Project Gets Initial Federal Funds," *Los Angeles Times*, February 25, 1995, https://www.latimes.com/archives/la-xpm-1995-02-25-me-36033-story.html.

33. Harrison Sheppard and Michael Coit, "Skepticism Greets LA's Toilet-to-Tap Water Project," *Los Angeles Daily News*, April 16, 2000.

34. Annette Kondo, "Sewer Water Reclamation Plan Comes Under Fire," *Los Angeles Times*, June 6, 2000.

35. "Safe Water, Iffy Politics," unsigned editorial, *Los Angeles Times*, June 18, 2000, https://www.latimes.com/archives/la-xpm-2000-jun-18-me-42193-story.html.

36. Joel Wachs, "'Toilet-to-Tap' Questions Must Be Answered," *Los Angeles Daily News*, May 15, 2000.

37. Marc B. Haefele and Anna Sklar, "Revisiting 'Toilet to Tap,'" *Los Angeles Times*, August 26, 2007, https://www.latimes.com/opinion/la-op-haefele26aug26-story.html.

38. Marjie Lundstrom, "L.A. Residents Aren't Lapping Up Toilet-to-Tap Water Plan," *Sacramento Bee*, April 20, 2000.

39. Haefele and Sklar, "Revisiting 'Toilet to Tap.'"

40. CH:CDM, A Joint Venture, "City of Los Angeles Integrated Resource Plan: Facilities Plan, vol. 2: Water Management," July 2004, 6-3, accessed January 20, 2023, https://www.lacitysan.org/cs/groups/public/documents/document/y250/mdew/~edisp/cnt010377.pdf. Mia Rose-Wong with the Los Angeles Department of Water and Power said that in 2022 the East Valley Project was revived by the City of Los Angeles and rebranded as the Groundwater Replenishment Project (GWR). GWR is slated to be replenishing the San Fernando Groundwater Basin by 2027 using

reverse-osmosis-treated recycled water. Email correspondence with author, May 23, 2023.

41. Hector Becerra and Andrew Blankstein, "Southland at the Tinder Mercy of a Record-Breaking Dry Spell," *Los Angeles Times*, June 30, 2007.

42. Haley Smith, "California Suffering through Driest Years Ever Recorded, with No Relief in Sight," *Los Angeles Times*, October 3, 2022.

43. Bethania Palma-Markus, "Water Board Considering Proposed Reclamation Plant," *San Gabriel Valley Tribune*, September 17, 2008.

44. Kerry Cavanaugh, "Mayor Revisiting Toilet-to-Tap Plan?," *Los Angeles Daily News*, May 15, 2008, https://www.dailynews.com/2008/05/15/mayor-revising-toilet-to-tap-plan/.

45. Cavanaugh, "Mayor Revisiting Toilet-to-Tap Plan?"

46. "L.A. May Flush Old Fears of Toilet to Tap," unsigned article, *Los Angeles Daily News*, June 22, 2008, https://www.dailynews.com/2008/06/22/la-may-flush-old-fears-of-toilet-to-tap/.

47. Palma-Markus, "Water Board Considering Proposed Reclamation Plant."

48. Dana Bartholomew, "DWP Testing a Way to Recycle City's Wastewater into Drinking Water," *Los Angeles Daily News*, June 14, 2011, https://www.dailynews.com/2011/06/14/dwp-testing-a-way-to-recycle-citys-wastewater-into-drinking-water/. See also City of Los Angeles, "One Water LA, Stakeholder Workshop #2 (Phase 2)," June 29, 2016, Meeting Notes, 4, https://www.lacitysan.org/cs/groups/public/documents/document/y250/mdey/~edisp/cnt012731.pdf.

11. Pure Water SoCal and Operation NEXT

1. California Department of Water Resources, "Producing and Consuming Power," accessed February 17, 2023, https://water.ca.gov/What-We-Do/Power.

2. Metropolitan Water District of Southern California, "Metropolitan Board Awards $13.9 Million Contract to Construct Recycled Water Demonstration Facility," July 12, 2017, https://www.mwdh2o.com/media/18082/rrwp_boardapproval_release.pdf.

3. Deven Upadhyay, executive officer and assistant general manager, Metropolitan Water District of Southern California, interview with author, May 17, 2023.

4. Rupam Soni, community relations team manager, Metropolitan Water

District of Southern California, interview with author, November 30, 2022.

5. "The Sorcerer's Apprentice," unsigned article, *Art and Popular Culture*, April 6, 2012, http://www.artandpopularculture.com/The_Sorcerer%27s _Apprentice.

12. Water Diversion, or Water Reuse?

1. A. Park Williams, Benjamin I. Cook, and Jason E. Smerdon, "Rapid Intensification of the Emerging Southwestern North American Mega- drought in 2020–2021," *Nature Climate Change*, 12, February 14, 2022, 232–34.

2. Camille C. Touton, commissioner, US Bureau of Reclamation, presen- tation, Colorado River Water Users Association Conference, Las Vegas, December 16, 2022.

3. PRISM, "Colorado River Compact, Signed in Santa Fe, New Mexico," November 24, 1922, https://prism.lib.asu.edu/items/62719.

4. John Fleck, Erik Kuhn, and Jack Schmidt, "Does the Upper Colorado River Basin Routinely Take Shortages in Dry Years?," Inkstain.net, accessed March 8, 2023.

5. Scott Dance, "Lake Powell Is Rising More than a Foot a Day. But Megadrought's Effects Will Still Be Felt," *Washington Post*, May 11, 2023, https://www.washingtonpost.com/climate-environment/2023/05/11/lake -powell-water-levels-rising-drought/.

6. Anumita Kaur, "Changes Needed to Save Second-Largest U.S. Reservoir, Experts Say," *Washington Post*, February 18, 2023, https://www.washing tonpost.com/climate-environment/2023/02/18/changes-needed-save -second-largest-us-reservoir-experts-say/.

7. Central Arizona Project, "Terry Goddard," accessed March 8, 2023, https://www.cap-az.com/board/board-members/terry-goddard/.

8. Central Arizona Project, "Explaining CAP's Federal Repayment: A Cheat Sheet," accessed March 9, 2023, https://knowyourwaternews.com /explaining-caps-federal-repayment-a-cheat-sheet/.

9. US Bureau of Reclamation, "Central Arizona Project," accessed January 15, 2023, https://www.usbr.gov/projects/index.php?id=504.

10. Central Arizona Project, "Water: Brought to You by Central Arizona Project," accessed January 16, 2023, https://www.cap-az.com.

11. The Central Arizona Project website says that its "average annual evaporation loss during a normal, non-shortage year is approximately 4.5 percent, or 16,000 acre feet from the aqueduct and 50,000 acre feet from Lake Pleasant. Seepage losses are 0.6 percent, or 9,000 acre feet." That's a total of 75,000 acre-feet, which converts to more than 24.4 billion gallons. See Central Arizona Project, "Know Your Water News, CAP System Loss: Seepage and Evaporation," accessed March 5, 2023, https://knowyour waternews.com/cap-system-loss-seepage-evaporation/.

12. V. L. McGuire, "Water-Level Changes in the High Plains Aquifer, 1980– 1999," US Geological Survey, last modified November 29, 2016, https:// pubs.usgs.gov/fs/2001/fs-029-01/.

13. World Population Review, "Flattest States 2023," accessed March 2, 2023, https://worldpopulationreview.com/state-rankings/flattest-states#.

14. Jennifer E. Zuniga, "The Central Arizona Project," US Bureau of Reclamation, 2000, 45, accessed April 6, 2023, https://www.usbr.gov/projects /pdf.php?id=94.

15. Zuniga, "The Central Arizona Project," 45.

16. Zuniga, "The Central Arizona Project," 23–24.

17. Zuniga, "The Central Arizona Project," 34–35.

18. Central Arizona Project, "Shortage Impacts," accessed March 9, 2023, https://www.cap-az.com/water/cap-system/planning-and-processes /shortage-impacts/.

19. Abrahm Lustgarten and Naveena Sadasivam, "How Federal Dollars Are Financing the Water Crisis in the West," *ProPublica*, May 27, 2015, https://projects.propublica.org/killing-the-colorado/story/arizona-cotton -drought-crisis/.

20. Circle of Blue, "Stranded Assets," accessed April 30, 2023, https://www .circleofblue.org/stranded-assets/.

21. Nick Walter, Central Arizona Project, email correspondence with author, May 25, 2023.

22. James Leggate, "Arizona Advances $5.5B Mexico Desalination Plant Proposal," *ENRSouthwest*, accessed April 2, 2023, https://www.enr.com /articles/55659-arizona-advances-55b-mexico-desalination-plant-proposal.

23. Glennon also sees a role for water markets, as well as more efficient water pricing.

24. Chenyang Hu, "Econometric Analysis of the Arizona Alfalfa Market," master's thesis, Department of Agricultural and Resource Economics, University of Arizona, 2019, accessed April 2, 2023, https://economics .arizona.edu/econometric-analysis-arizona-alfalfa-market.

25. Carly Becker, "The Value of Water," *Penn State Extension*, June 21, 2021, https://extension.psu.edu/the-value-of-water.

26. Dave Jones, "Cow Shower Study Examines Milk Production and Welfare Issues," *News UC Davis*, August 8, 2008, https://www.ucdavis.edu/news /cow-shower-study-examines-milk-production-and-welfare-issues.

27. Julie Murphree, "The Most Interesting Facts about Arizona Dairies You'll Ever Read," *Arizona Farm Bureau*, June 5, 2018, https://www.azfb.org /Article/The-Most-Interesting-Facts-about-Arizona-Dairies-Youll-Ever -Read.

28. Murphree, "The Most Interesting Facts."

29. Revol Greens, "About Us," accessed March 9, 2023, https://www.revol greens.com/about/.

30. Equilibrium, "Our Firm," accessed March 5, 2023, https://eq-cap.com /about-us/our-firm/.

31. Lisa Cownie, "Revol Greens," *Connect Business Magazine*, May 24, 2021, https://connectbiz.com/2021/05/revol-greens/.

32. Arizona Farm Bureau, "Yuma County," accessed January 27, 2023, https://www.azfb.org/about/counties/yuma.

33. *Daily News* staff, "Arizona Legislature Wants Feasibility Study for Long-Distance Pipeline to Replenish Colorado River Supply," *Mohave Valley Daily News*, May 11, 2021, https://mohavedailynews.com/news/131764 /arizona-legislature-wants-feasibility-study-for-long-distance-pipeline-to -replenish-colorado-river-supply/.

34. Tony Davis, "Once Again, Arizona Hopes to Import Out-of-State Water in the Face of Crisis," *Arizona Daily Star*, May 29, 2021, https://tucson .com/news/local/once-again-arizona-hopes-to-import-out-of-state-water -in-face-of-crisis/article_c47bf80a-be39-11eb-918b-13b88dd52f2f.html.

35. Brandon Loomis, "Pipelines? Desalination? Turf Removal? Arizona Commits $1B to Augment, Conserve Water Supplies," *Arizona Republic*, June 27, 2022, https://www.azcentral.com/story/news/local/arizona-environ ment/2022/06/27/arizona-lawmakers-bank-billion-dollars-augment-and -save-water/7736861001/.

36. Don Siefkes, "Suggestions for Smart Ways to Conserve Water," letter to the editor, *Denver Post*, May 15, 2022.

37. Don Siefkes, "We Could Fill Lake Powell in Less than a Year with an Aqueduct from Mississippi River," letter to the editor, *Desert Sun*, June 30, 2022, https://www.desertsun.com/story/opinion/readers/2022/06/30/we-can-lake-powell-less-than-year-via-mississippi-aqueduct/7751467001/.

38. Siefkes, "We Could Fill Lake Powell."

39. Janet Wilson, "A Pipe Dream, or a Possibility? Water Experts Debate 1,500-Mile Aqueduct from Cajun Country to Lake Powell," *Desert Sun*, August 15, 2022, https://www.usatoday.com/story/news/nation/2022/08/15/climate-change-west-mississippi-river-pipeline/10332092002/.

40. Wilson, "A Pipe Dream, or a Possibility?"

41. Roger C. Viadero, E. Dave Thomas, and Samuel Babatunde, "Meeting the Need for Water in the Lower Colorado River by Diverting Water from the Mississippi—A Practical Assessment of a Popular Proposal," Research-Gate, October 17, 2022, https://doi:10.13140/RG.2.2.21177.44640.

42. Don Siefkes, "It's Time for Army Corps of Engineers to Investigate the Feasibility of Moving Water West," *Desert Sun,* July 30, 2022.

43. 250,000 gallons per second = 33,420 cubic feet per second. The Colorado River's average flow is 22,000 feet per second. See "50 Largest Rivers in the United States (by Discharge)," unsigned article, worldlistmania, May 7, 2022, https://www.worldlistmania.com/50-largest-rivers-in-the-united-states-by-discharge/.

44. Viadero, "Meeting the Need for Water."

45. Wilson, "A Pipe Dream, or a Possibility?"

46. Paul Cofell, "If California Comes for Midwest Water, We Have Plenty of Dynamite in Minnesota," letter to the editor, *Desert Sun*, July 11, 2022, https://www.desertsun.com/story/opinion/readers/2022/07/11/if-california-comes-midwest-water-minnesota-has-plenty-dynamite/7827575001/.

47. Rod Rom, "The Drought-Parched West Wants to Take Mississippi River Water? Fat Chance! Or Is It?," letter to the editor, *Desert Sun*, June 26, 2022, https://www.desertsun.com/story/opinion/readers/2022/06/26/drought-parched-west-wants-take-mississippi-river-water-no-way/7708419001/.

48. Charles Babb, "Memo to the West on Water: Create Your Own Solutions

to Your Own Problems," letter to the editor, *Desert Sun*, July 11, 2022, https://www.desertsun.com/story/opinion/readers/2022/07/11/if-califor nia-comes-midwest-water-minnesota-has-plenty-dynamite/7827575001/.

49. Brandon Loomis, "As the Colorado River Shrinks, Arizona Looks at Water Recycling, Desalination, Taller Dams," *Arizona Republic*, December 14, 2022, https://www.azcentral.com/in-depth/news/local/arizona-environ ment/2022/12/14/arizona-reaches-new-water-sources-dams-desalinization /69716586007/.

50. Peter Annin, *The Great Lakes Water Wars*, 2nd ed. (Washington, DC: Island Press, 2018), 62–83, 317–18.

51. Henry Brean, "Mississippi May Help Ease West Drought, Mulroy Tells Chamber," *Las Vegas Review-Journal*, July 20, 2011, https://www.review journal.com/news/mississippi-may-help-ease-west-drought-mulroy-tells -chamber/.

52. US Bureau of Reclamation, "Colorado River Basin Water Supply and Demand Study," Environmental Resources/Reports, Lower Colorado Basin Region, Programs and Activities, accessed March 13, 2023, https:// www.usbr.gov/lc/region/programs/crbstudy/finalreport/Executive%20 Summary/Executive_Summary_FINAL_Dec2012.pdf.

53. Annin, *The Great Lakes Water Wars*, 62–64. See also Benjamin Forest and Patrick Forest, "Engineering the North American Waterscape: The High Modernist Mapping of Continental Water Transfer Projects," *Political Geography* 31 (2012): 167–83.

54. Annin, *The Great Lakes Water Wars*, 65–68.

55. The Water Resources Development Act of 1986 also banned diversions from the Great Lakes without the unanimous approval of all eight Great Lakes governors. See Annin, *The Great Lakes Water Wars*, 81–83.

56. National Weather Service, "Mississippi River Flood History 1543–Pres-ent," New Orleans/Baton Rouge, accessed March 9, 2023, https://www .weather.gov/lix/ms_flood_history.

57. United Press International, "Jackson Opposes Diverting Columbia Water," *Chico-Enterprise Record*, January 15, 1965.

58. "$1.3-Billion Colorado River Project Bill Approved," *Congressional Quar-terly Almanac*, 1968, accessed March 18, 2023, https://library.cqpress .com/cqalmanac/document.php?id=cqal68-1284109.

59. US Environmental Protection Agency, *National Water Reuse Action Plan:*

Improving the Security, Sustainability, and Resilience of Our Nation's Water Resources (Washington, DC: US EPA, February 2020), 4.

60. Sharon Nappier, national program leader for water reuse, US Environmental Protection Agency, virtual presentation, WateReuse Association Annual Conference, March 24, 2021.

61. Andrew Farr, "Biden Signs Infrastructure Investment and Jobs Act with Funding Boosts for the Water Sector," *Water Finance and Management*, November 12, 2021, https://waterfm.com/house-passes-infrastructure-bill-including-funding-boosts-for-the-water-sector/.

62. City of New York City, Department of Environmental Protection, "New York City's Water Conservation and Reuse Grant Pilot Program," accessed March 6, 2023, https://www.nyc.gov/assets/dep/downloads/pdf/water/drinking-water/water-conservation-reuse-grant-presentation.pdf.

63. City of New York City, Department of Environmental Protection, "New York City's Water Conservation and Reuse Grant Pilot Program."

64. City of New York City, Department of Environmental Protection, "Water Reuse Fact Sheet," accessed March 6, 2023, https://www.nyc.gov/assets/dep/downloads/pdf/water/drinking-water/water-reuse-fact-sheet.pdf.

65. Winnie Hu, "Please Don't Flush the Toilet. It's Raining," *New York Times*, March 2, 2018, https://www.nytimes.com/2018/03/02/nyregion/new-york-reduce-water-use-in-rainstorms-flush.html.

66. Clean Water Services, "Water Reuse," accessed March 6, 2023, https://cleanwaterservices.org/our-water/resource-recovery/reuse/.

67. Kristian Foden-Vencil, "John Day Leaders Hope Fresh, Local Vegetables Can Help Save Their City," *Oregon Public Broadcasting*, April 19, 2022, https://www.opb.org/article/2022/04/19/john-day-oregon-grant-county-leaders-greenhouse-project-to-lure-families/.

68. Natural Systems Utilities, "Gillette Stadium and Patriot Place," accessed March 6, 2023, https://nsuwater.com/portfolio-item/gillette-stadium-and-patriot-place/.

69. Tangent Company "Water Recycling Solutions—WaterCycle®," accessed March 9, 2023, https://www.tangentcompany.com/?page_id=12.

70. Wade Goodwyn, "Recycled Water Quenches San Antonio's Thirst," NPR, October 1, 2011.

71. US Climate Resilience Toolkit, "Water Recycling in Clayton County,

Georgia," accessed February 11, 2023, https://toolkit.climate.gov/case-studies/water-recycling-clayton-county-georgia.

72. Curt Yeomans, "Gwinnett Considered a Model for Treating, Reusing Water," *Gwinnett Daily Post*, May 8, 2015.

73. Jeremy Schiller, Grassland Dairy Products, presentation, American Membrane Technology Association, October 19, 2020.

74. Caroline E. Scruggs, Claudia B. Pratesi, and John R. Fleck, "Direct Potable Water Reuse in Five Arid Inland Communities: An Analysis of Factors Influencing Public Acceptance," *Journal of Environmental Planning and Management* 63 (October 2019): 1470–1500, https://doi.org/10.1080/09640568.2019.1671815.

Epilogue

1. US Department of the Interior, "Gifford Pinchot: A Legacy of Conservation," accessed March 7, 2023, https://www.doi.gov/blog/gifford-pinchot-legacy-conservation.

2. Aldo Leopold Foundation, "The Sand County Almanac," accessed May 10, 2023, https://www.aldoleopold.org/about/aldo-leopold/sand-county-almanac/.

3. Aldo Leopold, *A Sand County Almanac, and Sketches Here and There* (New York: Oxford University Press, 1949): 141–49.

About the Author

PETER ANNIN, A VETERAN CONFLICT and environmental journalist, spent more than a decade reporting on a wide variety of issues for *Newsweek*. For years he specialized in covering domestic terrorism and other breaking news, but he wrote extensively about the environment as well.

Since 2000 his writing has focused exclusively on water. In 2006 he authored *The Great Lakes Water Wars*, an award-winning book considered to be the definitive work on the Great Lakes water diversion controversy. In 2018 Annin published an extensively revised edition of *Water Wars*, which was recognized with an Award for Communications Excellence from the Great Lakes Protection Fund.

One of the few national water journalists in the United States, Annin regularly publishes op-eds in outlets such as the *Chicago Tribune*, the *New York Times*, and the *Washington Post*. He has a bachelor's degree in journalism from the University of Wisconsin and a master's in international affairs from Columbia University in New York. He directs the Mary Griggs Burke Center for Freshwater Innovation at Northland College in Ashland, Wisconsin.

Index